600ppm

A Novel of Climate Change

600ppm

A Novel of Climate Change

Clarke W. Owens

COSMIC
EGG
BOOKS

Winchester, UK
Washington, USA

First published by Cosmic Egg Books, 2015
Cosmic Egg Books is an imprint of John Hunt Publishing Ltd., Laurel House, Station Approach,
Alresford, Hants, SO24 9JH, UK
office1@jhpbooks.net
www.johnhuntpublishing.com

For distributor details and how to order please visit the 'Ordering' section on our website.

Text copyright: Clarke W. Owens 2014

ISBN: 978 1 78279 992 4
Library of Congress Control Number: 2014958365

A CIP catalogue record for this book is available from the British Library.

Design: Lee Nash

Printed and bound by CPI Group (UK) Ltd, Croydon, CR0 4YY, UK

We operate a distinctive and ethical publishing philosophy in all
areas of our business, from our global network of authors to
production and worldwide distribution.

CONTENTS

The persons, places, and events depicted in this book are fictional. Any resemblance to actual persons, places, and events is unlikely and wholly coincidental.

Clarke W. Owens

This book, and everything else, is for D.

"With carbon dioxide levels already at 385 parts per million (over 100ppm higher today than they were before the industrial revolution), experts agree that levels will continue to rise to between 600ppm and 1,000ppm by 2100."
—John Abatzoglou, et al, in DiMento & Doughman, eds., Climate Change (MIT Press 2007): 46.

The measurement of global concentration of carbon dioxide in the atmosphere for October, 2014, according to the Mauna Loa Observatory in Hawaii was 395.93. "The upper safety limit for atmospheric carbon dioxide is 350 ppm."
—www.co2now.org

1

A Year for Raspberries

This is a dark time. It is dark most of the day long. They say it was not always so dark all day long. They say once you could see through the black clouds. They say once there was a break in the lightning overhead.

Older folks say these things, and I've seen vids about it, but it's not like we never have any breaks from the dark and the storms.

I've seen the patches. Sometimes it's gray. Sometimes it's pale blue.

It blows my mind when it's pale blue.

My name is Jeff Claymarker. I was born during the Deluge of '25. They say the Deluge was the worst in a long string of disasters. It's on the vids. I've watched the whole thing on my wearable media.

And they say once there were more cities ...but I shouldn't overdo it with that phrasing. I know all about the coastal cities — Miami, New Orleans, New York. People make jokes about how they could have moved New York City to New Jersey and called it New New York, but in reality, when the sea level rose and swamped Manhattan, the people dug in. They filled in the lower floors and brought in landfill. They turned Manhattan into Venice. They were going to keep it where it was as long as they could. That's the way people are. So it's still there, and you can see it in a gondola.

But Miami and New Orleans are just plain gone. There are old people alive now who say they used to live there. The grandparents of the refugees.

The refugees come from all over, and they're not sure where they're going. In Ohio, where I lived until recently, we had many refugees from the South and West. The South was hot. The West was a desert, burning up in wild fires, no water anywhere—whereas, east of the Mississippi, we had an excess of rain, but we still had to ration water.

The rationing began as a way to equalize access in the western, desert regions, like California and Arizona, but we also had problems there in the wet areas with pollution from algae blooms and fracking contamination.

I have much more to say about water. But if I'm to begin at the beginning, I guess I'll start with the one-hundred-year-old man. Because not all the stories about "how it used to be" came from the vids and the refugees and their grandparents. Some of them were live performances.

My friend Tom and I went to see the one-hundred-year-old man at Tent City after gruel one evening. He was a performer among the homeless. He kept them entertained, and everything in the vids he said he had in his head, and then he would tell you.

Tom is my best friend. He's two years younger than me, and he's been to college—well, for six months, anyway. He said he studied general education there. He said he was doing OK—getting B's and C's until his tuition money ran out.

Almost nobody goes to college these days.

I've never been myself. I graduated high school, which is farther than most people get. I've tried to educate myself beyond my K through 12 learning. I've read tons of books, and have downloaded course outlines from the net and plowed through the reading lists. Tom says I'm at least as educated as he is.

Tom has a woman and a kid. His woman is Irene. His kid is Tom, Jr. Tom works through the same temporary agency that I do, and Irene does some on-call work as a nurse's aide, but none of us works permanent full-time. We're like most people, or were. When both Tom and Irene are working, Cassandra Lurie, a friend of ours, looks after little Tommy. Cassandra is married to Grady.

Grady Lurie does dry walling for his father, or so he says, but most of the time he is home taking care of their baby, Leah, while Cassandra works at a Stop 'N Go. When Grady is dry walling, Irene looks after the kids.

So we all look after each other.

Tom is about five feet nine inches tall and has curly brown hair and very thin facial hair. He's slender, like a blade, and he seldom claims to be hungry. Irene is about five two, with short dark hair, and is a little heavier than the average person, but is not obese.

Nobody is obese these days.

A terrible thing happened to Tom, and it began that night that we went to Tent City.

In the beginning of this story I'm going to tell you, Irene didn't like Tom hanging out with me. She used to tell him that I was worthless, shiftless, a nobody. She changed her view later on, but that's what she said in the beginning.

Well, I might have been worthless, but I had fun. That's why Tom hung around with me, because I found adventures.

There are some I wish I hadn't found.

So, like I said, we went to Tent City. The people were festive, talking, laughing, toasting marshmallows on sticks over trash fires. The tents were all shapes, sizes, colors, some with flaps open and families inside. You could see them in there, gathered, doing nothing. Overhead, the skies were gray and darkening, but there was no rain yet.

And down near the fences we saw cops questioning a T-shirted boy — caught in the act of possession. When we got closer, we saw it was raspberries.

Possession of raspberries was illegal. You could get a year in the pen for raspberries.

2

Robbed of Water Credits

The centenarian came out at dinner time because he took donations. He was mostly bald, but had a tail of long white hair draping his neck and back—a Ben Franklin kind of look, but much thinner.

Everyone's thin now. The historians say that the body types are similar to those in the Great Depression of the 1930s, but not like early this century. They say that early in this century people were fat, loved sugar and salt, and drank twenty-ounce soft drinks from plastic bottles. They talk about Coca-Cola. They say it was a great icon of a beverage, but only one of many.

There's not as many soft drinks today, I'm told.

There's not as much of anything as there was then, I'm told.

So this is what happened when Tom and I went to see the One-Hundred-Year-Old Man. This was an adventure I wish hadn't happened. This was how it all got started.

I asked Tom, "How many water credits you got?"

"Lots," he says. "Nine or ten, I think."

He'd be robbed of them before the night was through.

3

Where He Saw Apples

The old guy knew how to control a crowd—call it an air of theatricality, call it stage presence. It's a power some people have.

He stood out in front of the tents, a thin, bowling-pin shape against the darkening sky, his bony, bare, spotty forearms sticking out of his green sweatshirt sleeves and waving as he talked, like a magician gesturing to supernatural beings in his employ.

"I'm a century old," he said. "When I was born, the president was Truman. Do you children know that name? He dropped the bomb on Japan in that great big war with Tojo and Hitler. You know who Hitler was, don't you?—a man who rose on resentment and lived on hate—and that's something I used to wish human beings would unlearn how to do, but I don't guess I'll ever get what I wish for. There's no time left for me, anyway. But you have tomorrow. What'll you do with it, children? What'll you do with that big fat gift, Tomorrow? Is it even a gift you're happy to have? Oh, how we used to love Tomorrow, way back when!

It was hanging out there, like ripe fruit—do you know what fruit is, children? Have you ever seen an apple?"

One little boy in the gathering crowd said, "I have!"

And as the people laughed, he ran back under his mother's arm, she held him to her billowing dress, a gray, cotton dress with green apples printed on it, and the people laughed, "That's where he's seen it!"

4

Irritated All the Time

The old man was bantering back and forth like that with the people. You could see it was their custom. And then he went into the spiel that people knew him for. He rolled up the sleeves of his threadbare green sweatshirt, hitched up the belt on his jeans, stood with his spotty brown arms raised to the rat-tail sky and called out:

"I'm a Marketed To Man. I've always been Marketed To, from the time I carried a lunch pail to school, to the time I drove a car to a crummy job and lived on sandwiches in a cockroach apartment, to the time I commuted to my career until I couldn't work anymore and went on the dole, till the time the dole dried up.

"A Marketed To Man, that's me. That's what they call a Demographic. I was a Demographic! When I was a little schoolboy, they tried to sell me toys, tried to sell me candy. When I was a young man, they tried to sell me cars, and when I was in my prime, they tried to sell me bigger cars. And when I was mature, they pitched investments, vacations, whatever a mature man might want. And when I couldn't poke my wife anymore, they tried to sell me pills to bring back my manhood— "

The people laughed and hooted, some of the men making obscene-looking gestures. The women pressed the children's faces against themselves and shook their heads, but everyone was laughing.

Somebody in the crowd said, "If you get an erection lasting more than four hours, rent yourself out as a coat rack!"

The old man laughed along with everyone else.

"And one day," said the wild old man, "they stopped marketing to me. They said, 'He's dead, we don't have to market to him anymore! He's a Dead Demographic.' You see before you, ladies and gents, a Dead Demographic.

"But I'm still walking around. I still get up in the morning, I still take a piss, I still got a couple of teeth I got to brush. See here?" prying his mouth open with a finger, making one of the women squeal with amused revulsion.

"You're a Boomer!" someone called out, and his face lost its smile.

"Don't call me that," he said. "Call me a genius of survival! And if you want to know what a genius is, it's someone who's irritated all the time."

5

A Crime too Common to Report

Tom and I went out behind the fence afterward to smoke a Mary Jane. Once upon a time, Mary Jane was illegal, Tom says, and my mother says so, too. It's in the vids.

There was this cyclone fence, it's called—square wire mesh— and it was along the field out near the fairgrounds, where Tent City was located, on the main street leading into my old home town, Wooster, Ohio. Across from it was a gas station and some body shops—businesses dealing with a dying way of life—and if you followed the street eastward, you'd pass a brick building occupied by government folks, like the county recorder's office, the board of commissioners, the bureau of motor vehicles, and others. But out where we were, there wasn't much going on, unless maybe a cop drove by, or someone passed in a gas-powered car headed for the old highway.

You had to have a special license to drive a gas-powered car.

Tom and I liked that feeling of being on the edge of town, where nothing was happening. But it didn't work well for us, because that was where the robber found us.

The robber was, I guess, nineteen or twenty, thin-lipped, scrawny, nervous-looking, wearing an Encroachers T, black with skulls, and tattoos on his arms. He came snaking up along the fence line, his eyes on us the whole time, so I knew something was up, but I didn't know what.

He stopped in front of us, and I got a good look at his face. His hair was hay-colored and spiky, slicked with something to stay that way, so his head looked like a pool of water does when it's raining hard. His eyes were pale blue, crazy-looking, sunk in a reddish face, staring out at us like someone who'd been on a distant planet for a long time and might want to come home, but wasn't sure how to get there.

Tom and I took a puff and waited for him to say something.

"You got a water card?" he asked.

I thought, is that all?

Tom said, "Are you dry, buddy?"

If you were dry, Tom would share with you. He'd share with anyone who needed it. Tom had a good heart.

The kid said nothing, but he nodded.

"Come on, we'll get you to a station," said Tom, and that was when the kid pulled a piece out of some pocket in his ratty clothes, and we were staring down a blued barrel.

"Give me the card," he demanded.

Tom fished his water card out of his wallet and handed it over without a word. The kid turned and ran.

Tom and I shook our heads.

We didn't report it—too common.

6

When All the Refugees Were Happy

My mother said all the California refugees were happy in the Deluge—stood out in the rain and many were washed away. They'd not seen it before, not coming down like that, day after day. They spoke of drought and parched land, thirst and rationing. But everything is rationed now—food, energy, light, water. We have our cards and credits, we have our Genetically Engineered Wafers, we have our gruel rations at the stations, we have our lists of permitted and non-permitted groceries, which change every month, and we must follow them month to month. And that was when—I mean, at the time of the Deluge, and after —they rebuilt New York, and they founded new cities on the coasts of Arkansas and Georgia to replace New Orleans and Miami.

And that was when my father ran away. I never knew him.

My mother said my father could never keep a job, but I said, "So what? Nobody can."

"It was different then," said my mother. "The economy collapsed after the Deluge, but even before that, your dad would never stay with a job. There was always somebody or something about the job that he didn't like. And he was an alcoholic."

I went through some sturm und drang about my father when I was twelve, thirteen years old. I wanted to find him. But my mother said, "You'll never find him." And she was right.

She said I'd be going through a lot of sturm und drang for a couple of years, because I was going through puberty. "But one day," she said, "you'll wake up, and you'll be fifteen or sixteen, and all the sturm und drang will be blown away, mostly."

I hated her for saying that at the time. But she was right.

So my father took off and left us at the time of the Deluge, and the economy crashed, and the water problems got worse and worse in the West, and the refugees started pouring northward, and people began shifting around uncomfortably in other ways too.

Like, for example, when the Clevelanders came south to where we were, to get away from the poison in the lake.

And the Youngstowners came west to get away from the frack poison in the water wells.

And the population quadrupled down near Wooster, where I lived.

And it's doubled since then.

7

Kind of Like My Girlfriend

Wooster, Ohio, population 160,000. Most of us lived in the government high rises and trailer parks. Tom had trailer #525 and I was in trailer #530 in the Mad Anthony Wayne Park, which was state owned.

After the Tent City visit we went back to #525.

Irene was not happy. She'd been with us at gruel rations, but then she'd come home with little Tom, Jr., who was now two.

"You going to stay out all night with Jimmy Dean here?" she said to Tom.

Jimmy Dean was a young movie star in the twentieth century two-dimensional films who died in a car crash. We've all seen him in the old vids. She thinks I look and act like him, but that's her fantasy.

"It isn't even late, Irene, I don't know what you're so shook about."

"Hey, I better go," I said. "Good-night, you two. Bye-bye, little Tommy!"

And I was out the door.

I thought I'd go see if Kareena Hecht was home.

She was kind of like my girlfriend.

8

Nothing but a Lizard

I mean, if you want to make a movie out of my life, you can cast some gorgeous actress as Kareena Hecht, but that wouldn't be the truth.

If you'd asked me how I felt about Kareena at the time, I would have said—and I cringe to admit it—that I liked skinny women with twiggy legs and small butts and big boobs, which Kareena had, but she also had a face only a mother could love, as far as homely. I would have said that her complexion was like oatmeal, and she wore these triangular eyeglasses that made her straw-colored eyes look tiny. I would have said that she had a beak-like nose and thin lips. I would have said that she had kind of nice, wild, teased beige hair down to the shoulder, though, on the plus side. But—I would have said—she looked like a bird. Kind of storky, you know?

Irene had heard me talk about her this way, and it ticked her off.

"You men think you're God's gift to women, don't you? You think you're so superior! Well, you're nothing but a lizard, Jeffrey Claymarker! Do you hear me? Am I coming in loud and clear?"

Yes, Irene usually came in loud and clear.

She was right too. It wasn't like I was some prize. I guess I would have said that it wasn't that I thought I was superior to Kareena, but that I liked to say the way it was, is all. I did not think of the two of us as Romeo and Juliet, that's all I would have said. It sounds shallow to me now, after what we've all been through.

So how, given my cavalier attitude at the time, did she get to be kind of like my girlfriend?

Here is how that happened.

9

A Slime Ball Thing to Say

I had gotten a job through a temporary agency called T-People, stacking pallets. And Kareena worked there. She used to come by every hour to count the pallets. And I could see she kind of liked me. I'd catch her looking me up and down, and then she'd blush and act like she hadn't been, you know.

So I started flirting with her. I don't know why. I wasn't that attracted to her, but I had nobody else, and it got pretty lonely being twenty-five and healthy and alone.

So what would I say to her? I'd say, "Did you count all my pallets?"

And she'd say, "Yes, I did," in this tone like—only a moron would not realize that I did. It would be like, "Yes, I di-iiid." And then I'd say, "Well, I want to make sure you notice all my pallets. Because I stack them just for you to notice."

Stupid stuff like that, you know. Flirting is a lot of suds on the top of the beer stein, if you know what I'm saying.

So then we started going out. And we'd come back to her place and smoke crack, or snort a line, because Kareena was always into drugs of one sort or another. I didn't much care for it, but if she wanted to snort or smoke, we'd snort or smoke, and then we'd end up in bed.

I was up front with her. "I'm not looking for long-term," I said.

I told Tom I'd said this, over at his place one time, and Irene overheard it.

"That is a slime ball thing to say to a woman," said Irene.

Irene at that time was not my biggest fan, I guess you can tell.

Anyway, that job stacking pallets stopped after a few months, but I hung on to Kareena's phone number, and she didn't seem to mind.

So that's how that got started.

10

Stations

So anyway, I headed on over to Kareena's place. She lived in a government high rise on the west side. I carried my umbrella to keep from getting totally soaked, but the rain was not too heavy, more off-and-on.

Kareena's place was four stations west of our trailer park, not counting the station at the bottom of her high-rise.

I'd better explain about stations, because I don't know who will be reading this. I'm just writing it down and it will be like the old message in the bottle, I guess. And maybe whoever ends up reading it will live in a completely different society and won't know what it was like for folks in Ohio in 2051.

Stations are privately owned, but they're everywhere. They're where you can get water. You use your water cards. You can get water cards from the government, or you can buy them at stores, but you can only use them at stations.

Stations are usually pretty crammed with people because access to water in your house is only at certain hours of the day. You get three hours in the morning, two hours at dinner time, and two hours at night, from nine-thirty to eleven-thirty. If you need water at any other time, you've got to go to a station to get it.

So there are stations all over the place, and they do brisk business. The hard thing is to get the cards with the credits on them. If you don't have a job (and lots of people didn't, and don't) the only way to get water credits is through government subsidies, and you only get so many of those per month.

As you can imagine, the cards are pretty in demand. People steal them and sell them on the black market.

Like that kid who held up Tom on the way to Tent City.

Yeah. That was only the beginning of the story.

But I was on the way to Kareena's high rise, so that's where we'll go next.

11

Out of Minutes

So I pass the four stations and I get to the station at the bottom of Kareena's high rise, and I pass that, and go on into the lobby. I go down to the second bank of intercoms and look for Kareena's number, which is 22-A, because she's on the twenty-second floor.

I run my fingertip over the detector light.

I get the recording.

Hi. This is Kareena. I'm not home. Leave a message if you're so inclined.

"Hey," I say. "Kareena, this is Jeff. Just thought I'd stop by 'n see if you were home. No luck, I guess. Should've called first but I was out of minutes. I'll try again pretty soon. Later."

So I decide to call it a night, and I go home.

12

We Blew It In 1861

As far as water.

The problem in the eastern half of the US of A is not quantity. I mean, it rains all the time. The heat holds more moisture, they say. On the vids they say it didn't used to rain so much in the East, and it wasn't always so dry in the West, though it was always dryer there than here.

The problem is not the water tables per se. It's the contamination.

Contamination from algae blooms. Contamination from phosphorus run-off. Contamination from frack wells. Then, too, as far as quantity, there was prior depletion of the tables from massive water use by frackers, back before it was made illegal. Which was mostly after the wells were played out. I am talking about four years ago.

So the laws were all for show, they say. They were for people who were ticked off at the oil and gas industry, they say.

"Easy to be ticked off at the oil and gas industry after the wells are played out," said Wesley Wright on the Patriot Network whom I used to listen to at the time.

Then, as far as water supply, you've got the Water Transport Act, which was supposed to help out the West.

We practically went to war over this, East versus West, in the US of A. The federal troops had to be brought in.

There is no way you can beat the federal troops, even now, when they're only a pale ghost of what they used to be—which is what I heard on the Patriot Network at the time.

At that time, Wesley Wright was my favorite pundit on the PN. He said if it weren't for the power of the federal troops, and the money in the Defense Department, the country would have split up into autonomous regions by now, if not into fifty independent states, certainly after the Deluge of '25.

"You'll never do it now," said Wesley Wright. "We had our chance in 1861. And we blew it."

13

The Worst Luck We Ever Had

It's funny to think of it now, but at that time, in 2051, I considered Wesley Wright to be really glue. He was only like thirty-five or forty, something like that, yet he had totally white hair. He combed it back and sprayed it tight in a big pompadour, and he wore string ties with his blazers, like a country western singer from the twentieth century. He had these huge, bright, staring, yellow-colored eyes, which I didn't know whether they did that with special effects or what. And he had this powerful Roman nose and this florid, high-cheek-boned face. So when he looked into the camera, you felt like he had you fixed in that weird, wide-eyed, hawk like gaze.

And he told it like it was, I used to think.

"We'd be so much better off as fifty countries, or even five countries," he said.

"Face it, Friends: "he always called his viewers "friends"" people in Vermont have got zilch in common with people in Texas. People in Seattle or San Francisco have got zippo in common with people in Montgomery, Alabama or Atlanta, Georgia. So, in any national election, you've got half the country miserable no matter who wins. I'm telling you, the worst luck we ever had in this country was named Abraham Lincoln. That's a simple fact."

He always said that. "That's a simple fact."

He ended up running for president. More about that later.

14

Hypnotica

One night, not long after Tom and I went to Tent City and Tom got robbed of his water credits, and after I got some minutes on my phone, I got this frantic call from Irene, of all people. It was one day after there was this terrible news story in the vids about a young local girl who'd been murdered and her body was found in a dumpster, cut into four pieces.

"Jeff! The police came and picked up Tom! They took him down to police headquarters. I don't know what it's all about. We've got to go get him out!"

"Wait a minute, slow down, Irene. Are you saying they arrested him?"

"Yes! No. I don't know. They said they wanted to talk to him. They asked him to come down to police HQ, and he said OK, and they gave him a ride. I didn't go because I didn't want to take little Tommy down there and I had no one to look after him. Jeff, I'm scared to death! I think it has to do with that girl that was murdered."

"How do you know that, Irene?"

"They said something. They asked if he knew that girl. She had a weird name, I remembered it from the news story, but then I forgot it."

I thought for a second.

"Hypnotica," I said.

"Hypnotica! That was it! That was how I knew it was about her, because she had that strange name, and I heard them say it. 'Did you know this girl, Hypnotica,' they were saying."

"I don't think he knew her, did he?"

"No, of course not. We don't know any girls who get murdered and cut in pieces! Who hangs around with people like that? Jeff, we've got to go down to the police station and get Tom out before they lock him up."

"Well, did they arrest him or not? Did they take him away in handcuffs?"

"No, they just asked if he would come down and—oh, how did they say it? 'Are you willing to do a voluntary interview,' they said."

"That doesn't sound like an arrest to me."

"Jeff, I'm scared. I found a neighbor to look after Tommy but I don't want to go to police HQ by myself. They might lock me up, too. Will you come with me?"

"I'll come, but if they want to lock you up, I won't be much good," I said.

Well, I went with her.

15

Neither One Noticed

We went through the glass doors at the Justice Center and up to the window where the women were sitting, like at a ticket window.

"We came to see Tom Glendinning," we said. "He's being interviewed by the police."

"Did the police ask you to come?" said the women.

"No."

"What's your name?"

We gave the women our names.

They pushed some buttons on their control panel and activated their speakers. They told the police about us. The police said, "We're not through talking to him. Tell them to wait."

"You'll have to wait," said the women.

We sat down on the benches in the waiting room. You couldn't go anywhere else, because all the doors were electronically locked, unless we wanted to go back where we came from.

We waited for two hours almost. Then we heard the electronic bolt slap open in the glass door behind us and Tom came out into the waiting area.

Irene was all over him.

There was a cop behind him. When he saw me, he said, "Who are you?"

I told him.

"I'd like to ask you some questions, Claymarker. You got a few minutes?"

"OK," I said.

Tom had his hands full with Irene, who was an emotional wreck. I don't think either one of them noticed when I went back through the door and down the hall with the cop.

25

16

I Thought That Would Be the End of It

"It's not enough to hold the gun and pull the trigger. You've got to direct the bullet."
—Wesley Wright.

The electronic bolt went whack behind me, as I followed the officer down the hallway to some rooms near the jail.

The officer was of medium height, balding, fair-skinned, somewhat heavyset. He walked the way police officers walk: like they're used to occupying space. Like they've got control of that space, and you'd better not challenge them on it.

I didn't want to challenge anybody.

"I'm gonna go get the detective," said the officer, as he led me into one of the rooms and had me sit in a green plastic chair next to a bare brown table.

Then he left.

It was the barest of rooms, with a smoked glass window next to the door. Above the desk was a camera aimed down at me. So I knew I'd be digitalized.

I was sitting there for five or ten minutes, when a plainclothes guy came in. He was taller than the other officer, not as heavyset, middle-aged, dark hair, glasses.

"I'm Detective Hayes," he said. "You're Jeff Claymarker?"

"Yes, sir."

He sat down in the one other chair the room held, across the table from me.

"Why we wanted to talk to you, Jeff, is Tom told us you're a friend of his, is that right?"

"Yes, sir."

"Well, look, you're not under arrest, OK? You can leave any

26

time you want. So if you stay and answer questions, it's voluntary on your part, you understand?"

"Yes, sir."

He put a piece of paper in front of me and a pencil. The paper said the same thing, about me being there voluntarily, and he asked me to sign it. I did.

"This is about that girl who was killed, Jeff. You know about that, don't you?"

"Yes, I saw it on the vids. Hypnotica."

"That's right. We wondered if maybe you knew more about it than just what's on the vids. Because, I'll tell you the truth, we found a water card on the body that belonged to Tom."

"Ohhh," I said. So that was it! "That water card—a guy stole it off of Tom when I was with him."

The plainclothes guy—Detective Hayes—brightened up. "Well, that's what Tom said."

"It's the truth."

Hayes asked me to tell him what happened. He got a full description of the robber from me. I figured they knew I wasn't lying because they could match up what Tom said with what I said, and no one had telegraphed the questions to me.

At the end of it, the cop said, "You've been very helpful, Jeff. You can go now. If I have anymore questions, I'll be in touch. And you let me know if you see that guy again." Meaning the kid who stole the water card.

"I sure will," I said.

He seemed like he believed me, and I thought that would be the end of it.

17

I Can't Imagine a Peaceful World

"A patriot is someone who supports his country, right or wrong. This means you must be willing to do wrong when your country requires it."
—Wesley Wright

I went home and watched the streaming news. It was all about the wars. There was a lot of glue footage of the robotic bombers blowing up buildings and stuff. Then they'd have some guy on the ground interviewing families who'd been hit, and you'd feel bad.

I guess war is hell, but that's all we've ever had. It's us against the Caliphate, and the Caliphate against itself. The entire Middle East and northeastern Africa, in civil war all the time. Three to six or seven groups in each country, fighting it out at any given time. Tom said we should declare victory and go home, but Wesley Wright said we can't do that because there are people in these desert countries who hate freedom, and that's why they hate us, because we're free.

Which, if you think about it, is pretty weird. I mean, people sitting around with their tents and camels thinking, I really hate freedom. I think I'll kill some Americans, because they're free, and I hate that. I mean, that hardly makes any sense. Because if you think about a guy in prison or something, he's probably not thinking, gee, I hate people on the outside, because they're free. First chance I get, I'm going to kill someone on the outside because he's not in here with me.

I think they probably have reasons for hating people and wanting to kill them other than that.

But I don't have my own talk show.

The President came on the vids, and said, "We will track down the terrorists. We will find them in their lairs. We will bring justice to them."

And I'm wondering like, how long has this been going on?

If you listen to your grandparents and watch the history vids, it was like, once upon a time there were only two great super-powers, the U.S. and the Soviet Russians, and they never went to war because they were scared to death they'd blow each other up with nuclear weapons.

I can't even imagine such a peaceful world as that, now.

That sounds like heaven.

18

Not Good for the Jobs Situation

My favorite channel was Wild Beast World. They had vids of all these crazy animals, like tigers and panda bears and polar bears and great apes and elephants. I wish we had animals like that now. I'd go see them, if we had them in zoos.

The apes were really amazing. They were like people. They used to have these apes called chimpanzees that could understand sign language. They'd sign at you and say, "Koko wants a drink of juice. Koko wants a banana."

That is so glue. I wish we still had chimpanzees and I could see one live.

Elephants, god. They were the greatest, but they're all gone now, wiped out by ivory poachers. They say they were as smart as people, and much kinder. They would stand over the places where their relatives had been killed and weep tears. They would protect each other from lion attacks by grouping together, just like an army. They would take baths in lakes and have a great time.

I wish we had elephants still.

I wish we had lions, too.

Here's something I learned on Wild Beast World. There didn't used to be all these jobs for people hand-pollinating almond trees and fruit trees because it used to be done entirely by insects called honeybees.

Wow.

I wish we had some of those, so I could see how they did it.

But I guess that wouldn't be good for the jobs situation.

19

A Bowl of Mango Slices

Of course, even with hand-pollination, fruits are severely rationed and you can get jail time for buying black market fruit. Fruit addiction is a serious problem, though. It seems you can't stop some people from getting their hands on it, crime or no crime. Of course, if you've got enough money, you can get a special permit to buy fruit from a government store. I guess there's no fruit shortage for the rich. For the rest of us, it's black market or nothing.

One night Tom and Irene took me and Kareena to a speakeasy for some black market fruit.

The term "speakeasy" comes from Prohibition days, it says on the vids. It was back in the 1920s when it was illegal to possess alcohol. So they had these bars where you went to the door and gave a password, and they let you in and you could buy liquor.

We have the exact same thing, only instead of liquor, you buy apricots or bananas or peaches or plums. They serve you banana splits or other ice cream dishes with strawberries or cherries or raspberries or blueberries in them. They sell mango juice and pomegranate juice and papaya nectar—stuff you can never get in a non-government store at a reasonable price.

The place we went to was called Uncle Donald's. It was on the east end of town, near the antique railroad tracks, in the back of a two-storey house. You went around the side of the house, and through a low fence. There were a bunch of trees close together. You had to brush the low-hanging branches aside to get to the door in back, which was dark blue. Tom gave a special rap on the door—Rap. Rap. Rap-rap-rap.

A little square peep hole opened up at eye level, about six inches wide, and a pair of blue eyes stared out.

"Where you from?" said a voice attached to the eyes.

"Strawberry fields," said Tom.

Strawberry fields was the password. It was based on some old song.

We heard the door unlock and then it opened with a shudder. It was a pretty flimsy-looking door, overall.

The blue eyes we'd seen through the peep hole belonged to a grinning, old guy, with a long, curly white and sand beard. He laughed when he let us in and you could see several gaps in his crooked teeth.

Behind him was a bar. There was a row of men sitting there, and we could see only their backs. They had plates in front of them and they were eating peach slices, grapefruit, all kinds of colorful fruit you'd never see in any store. The whole room had this kind of fresh, citrusy smell.

There were little tables closely spaced around the room and men and women hunched over them, talking in confidential tones or laughing at private jokes, all of them with some kind of fruit concoction, from berry cobblers to strawberry daiquiris. The whole place seemed so incredibly happy, compared to the gruel bars where we ate most of our meals, or to the state run carry outs where you bought fortified wafers made of approved ingredients.

In the history vids, you could learn that fruit was once legal. People used to eat it all the time, especially in places like California, where they once had these huge farms that cranked the stuff out like a factory. All those farms were owned by Standard Oil, they say.

That was a big oil company that is only into publishing vids now.

That was before the honeybees went extinct, which I mentioned before.

Honeybees were insects that used to live in hives and they would fly out and pollinate all kinds of fruit plant life, and the result was an abundance of fruit. When the bees went extinct, government and industry partnered to promote jobs for people to pollinate the plants by hand, and that's what we do now, but it's not anywhere near as efficient, they say. Consequently, fruits and fruit products are high-priced and many are only available with special permits. Without permits, the forbidden ones are illegal to possess.

But you can never stop people from getting their hands on something they really want, especially if it's something really great to eat and is good for you, like fruit is.

The cackling guy with the long curly beard showed us to a table near a fragrant potted plant. There were several of these plants scattered among the tables, and some small potted trees as well. The plant next to us had big, green, rubbery leaves, but I don't know what it was.

"So what do you guys want?" asked the bearded guy.

"Banana ice cream with blueberries and raspberries," said Kareena.

"A nectarine," I said. "One with lots of spots on it, and plum colored."

"Lime sherbet with a kiwi," said Irene.

"I don't know if I can remember all this," said our waiter, and then he decided to fumble in his apron for a virtual pad. "What about you, guy?" he prompted Tom.

"A bowl of mango slices," said Tom, with his mind already made up. "A bowl of mango slices."

He swore to us that mango slices were food for the gods. But we each had our favorites that we'd been saving up to enjoy for a long time. It had been almost five months since I'd had any contraband fruit, and back then it was a dried apricot. Five months earlier had been winter and nectarines had been out of season. I couldn't remember the last time I'd had one, but I knew it had been great, and I wanted one again.

We knew it was dangerous. We knew we could be raided at

any time. But we just had to have that fruit. And, as it turned out, problems with the law came in a different form, just a couple of days later.

20

Stupefied

Tom and I went to see Grady Lurie. He had a little one-seater wind vehicle. Grady lived in a quadplex in mid-town, not far from our trailer park, and he kept the WV in a little yard or parking lot behind the house. It didn't run, and he invited us to work on it, because Tom had taken a small motors class during his six months in college and he could fix almost anything. Most of the time.

Well, we worked on that WV for three hours, but we didn't have any luck with it, so we walked back to the trailer park down near the fairgrounds.

As we approached Tom's place, we saw a cop car in front.

Cops have got most of the only gasoline-powered cars still around. The gasoline comes from natural gas, and everyone knows that stuff is played out. It comes entirely from storage. That's why it's illegal for anyone else without a permit to use it, and that's why it's mostly cops who have gasoline-powered cars.

So you'd notice one of those, even if it wasn't black and white with WPD all over it.

As we got close to the vehicle, we saw the cop coming away from the front door. Irene was standing in the doorway, behind the cheap screen, crying, and behind her we could hear Tom, Jr. crying too. (Tom was only a toddler, just old enough to walk.)

The cop strode up to us. He had some kind of paper in his hand.

"Which one of you is Tom Glendinning?"

"I am," said Tom.

"You're under arrest for murder," said the cop. He had the handcuffs out.

"Murder? What are you talking about? I haven't murdered anyone!"

"Hands behind your back, son. I've got a warrant here."

Tom knew better than to resist.

"This is not that same girl, is it?" I said. "That Hypnotica one?"

"He'll get all that information at the right time," said the cop, shoving Tom into the back of his cruiser, behind the electron mesh.

Irene ran out to the car.

"Go back inside," said the cop.

"I just want to talk to him for a minute," said Irene, crying.

"Sorry. You can talk to him later."

"See if you can get me bailed—" Tom was calling out from the back seat when the door shut.

Then the police car drove away, and we stood there, stupefied.

21

Almost a Year's Pay

We could not get in to see Tom until the following Saturday. The cops let us in for half an hour each. Irene took the first half hour, then Grady and I split the second half-hour, fifteen minutes each. The cops wouldn't let any two of us in at the same time.

When I went in, I had to go through a door into a little room with windows and phones in front of each window. On either side of each window was a plastic chair. They had Tom sitting at one of these already. I sat across from him and we picked up the phones to talk, which was more confidential than using speakers.

"Thanks for coming in, man."

"You're welcome, man. Hey. What's it all about, Tom?"

"It's that same girl. Hypnotica Christiansen, her name is. Was. They say they have an eyewitness who saw me kill her and chop her up."

"But you never did that, Tom."

"Duh, tell me about it."

"So who is this witness?"

"They say his name is Guy Milteer. That's all I know. They didn't have a picture of him, so I don't know what he looks like. I think it's that same dude that held us up. It's got to be."

"Why would the police believe that punk?"

"That's what I don't know, Jeff."

"You've got to get a lawyer, Tom."

"They said they were going to appoint the Public Defender."

"Have you talked to them yet?"

"No. The cops said they'd come and see me at the appropriate time."

"Don't let them railroad you, Tom."

"No one's railroading me. I'm going to defend myself. I'm not guilty, Jeff!"

"I know you're not."

But even as I said that, I was thinking that sometimes people get convicted of things they didn't do, just like other people do things and never get convicted. I hoped nothing like that would happen to Tom.

As for bail, it was set too high. It was three hundred thousand dollars cash or surety, with no ten percent. It was like almost a year's pay for someone with a job.

22

These Whole Deep Layers Underneath

Some of the story was in the local vids, naturally. The way it went was this.

On the night that Tom and I went to Tent City to watch the One-Hundred-Year-Old man do his rap, this Guy Milteer character claimed he met Tom and me behind the fence.

The weird thing about this story is that this Milteer guy used my name and claimed to know me. It was in the vids. "He claimed he met Glendinning and a friend of his, named Jeff, as they were smoking marijuana behind the fence at Tent City."

This was weird, because the only guy we saw that night was the thief in the Encroachers black T-shirt with the skulls on it and the tattoos on his arms. And I didn't know that boy from Adam. Never saw him before in my life. And that character did not talk to us, did not ask our names, just pulled out a pistol and demanded Tom's water credits after Tom had already offered to get him to a station for free.

You see how weird this is. That guy had no reason to rob us for the water card. Oh, sure, he could have sold it on the black market for more than two or three credits worth, but how much profit would he have made? Tom had only like ten or twelve credits total to begin with.

So this supposed "witness," if it was the same guy, said that I left and he went with Tom to a bar, where they met this Hypnotica and offered her water credits in exchange for sex.

Now that part, right there, that was a total fabrication, because Tom didn't do stuff like that. He had no reason to use up his water credits trying to get sex, because he had Irene; and besides, he used his water credits for his family—little Tommy. He was very good about it. He didn't go around looking to sell his credits on the black market.

So the cops had a total liar as their main witness. But I didn't know how he got my name, or—if it was the hold-up man—why he held us up, or why the police believed this lunatic.

Anyway, the story was, after they took this Hypnotica back to some abandoned shack on the edge of town, that Tom went berserk and killed her, while this Milteer character tried to stop him, and then ran away when he couldn't. This Milteer supposedly panicked. Two days later some bum found the body buried in a ditch fifty feet from the shack. Then Milteer went to the police and told them what happened.

That's what was in the vids.

What I figure is that the police came to interview Tom, and then me, after they found the body, but before this Milteer came forward, because when they interviewed us, they didn't say anything about another suspect, much less a witness. And they were real eager to hear my story about the boy who robbed Tom, which they also said at the time Tom had told them.

So they should have known this Milteer was lying, because if he wasn't, why wouldn't he have come forward right after it happened?

So I was thinking, wow. What you see in the vids is just one tiny snippet of what goes on, and there are these whole deep layers underneath it that you never see.

23

Death Doesn't Hurt, Thought Hurts

The next thing that happens is, I get a call from the Public Defender's office and they want me to come in for an interview. They advise me I don't have to come in, and that I might want to consult a lawyer first, and that I have a right not to say anything to get myself in trouble. But the reason they want to talk to me is that they think I have information that could help Tom.

So of course I go in. I mean, I've got nothing to hide, right? And I want to help Tom.

The Public Defender's office is right next to the courthouse with the caryatids or whatever they're called, the Atlas statues, holding up the building. And I go through the glass door and through the wooden door and up to the window, and the lady tells me to wait, so I sit on the chair. Then out comes the Public Defender, which is a lady, very tall, about thirty-five years old, with light brown hair down to her shoulders and coffee colored skin, white blouse, black skirt.

Most people in America have coffee colored skin. Grady does. Tom and Irene and I are white. We're minorities. Wesley Wright is a minority white guy. In the vids you can learn that the people with the colored skins used to be the minority and they used to get treated badly, going back all the way to that wonderful American institution, slavery. In the twentieth century they used to lynch people just for being black. This is true, I tell you. I know it's unbelievable.

I'm glad things aren't like that now. Especially because I would be the one getting lynched.

"I'm Jade Bessemer," says the tall Public Defender lady, holding out her hand, which I shake. "Are you Jeff?"

"Yes, ma'am."

"Thanks for coming in, Jeff. Come on back with me to my office and let's talk."

She takes me through another door and down a corridor, past some other doorways with, looks like, other lawyers in their offices at their desks. Inside her office, she closes the door behind me and has me sit in a comfy chair with silver frame handles and big, soft, brown cushions, facing her desk. She's got her framed license to practice law on the wall and some book and vid cases behind her. On the shelves, though, she's got these little toy tractors and cows and barns and stuff, and I'm wondering what that's about. But I don't ask.

She looks at me with these intelligent green eyes, and she says, "Jeff, I understand you're Tom Glendinning's friend."

"His best friend as far as I know, ma'am."

"How long have you known him?"

"Since fourth grade, ma'am."

"Jeff, just call me Jade, OK?"

"OK, m—Jade."

She went through the whole spiel about how I didn't have to answer questions, and I might want an attorney first, and all that, but I told her I already knew that, and I wanted to answer questions if I could help Tom.

"And you'll tell me the truth, Jeff?"

"Absolutely."

"Well, then, do you remember back on April 4th—"

She was just like a lawyer, wanting to refer to dates and times, and get it exactly right. I remembered it was in April that Tom and I went to see the One-Hundred-Year-Old Man, but I didn't have the exact date. But she said Tom said it was the 4th, so I said, then that's what it must have been, and it sounded right.

And I told her what happened. How we went to smoke a Mary Jane out by the fence afterward, and this boy comes up wearing an Encroachers black T-shirt with skulls, and tattoos on his arms, and—

"Who are the Encroachers?" she asks.

She wanted to know everything.

"They're a Death Rock group."

"Death Rock?"

"Yeah. Like, you know the song, 'Death Is All Around'?"

"No, I don't know that one."

"What about 'Death Don't Hurt, Thought Hurts'?"

"I'm afraid I haven't heard that one either."

"I guess you don't follow music, huh?"

"Not all of it."

"Well, I'm not the biggest fan of Death Rock either, to tell you the truth. I'm more of a Collapse Rap guy. Collapse is so glue."

"Glue?"

"Yeah. Glue. You know, like Stick? Like Rapid? Like Chic? You know, it's positive, I like it. It's Glue."

"Ah."

So then I told her all about how this boy with the Encroachers T asked us if we had a water card, and how Tom had a water card, and how he offered to get the boy to a station to get him some liquid if he needed it. And then how the boy holds us up at gunpoint to get the water card, which made no sense, because he could have had two credits' worth for free, and Tom only had like ten credits on the card. So this guy is doing an armed robbery for eight credits on a water card.

She interrupts me at that point.

"Jeff. Tom says you didn't report the robbery to police."

"No, we didn't bother."

"Why not?"

"What for? For eight lousy water credits? Let him have 'em!"

"That's what you thought at the time?"

"Yeah. People get held up all the time for water cards. We said, 'Screw it.'"

Jade asked me all about everything we did that night, and everything I could remember for the several days afterward, including the interviews with the police. She was real interested in those interviews.

"They never mentioned anything to you about a witness to the murders?"

"No, ma'am, not—"

"Jade."

"Jade. No, not at the time. I only found out about that after they arrested Tom and put him in jail."

"Jeff, are you safe living where you are?"

"Sure, why wouldn't I be?"

"I want you to be very safe, Jeff, because you're our alibi witness."

She said something was fishy about what she called the "State's case," and I was going to be the star witness at trial. And we were going to win.

That's what she said.

24

You Heard It Here First

Every night was dark and rainy. The clouds were black. Behind them lightning lit up the sky like a huge pitchfork and cracked overhead and the tents in Tent City strained against their stakes, and some of them flew away. And the tin walls of our trailers shuffled up and down and rattled as the wind howled, and this was our summer, day after day.

The only fun I had was updating my wearable tech. Then I'd catch Wesley Wright on the Patriot Network.

Most of the time he was criticizing President Goya.

"Goya's got no Get Up and Go-ya," he said. "The day that it becomes impossible for the U.S. of A. to maintain three fronts against the Caliphates, that's the day the U.S. of A. slides into second-class status as a nation. Some people say it's already happened. I tell you, my friends, it may LOOK as if it's already happened, when you look at the weak, inept, sluggish foreign policy of this president. We're not there yet. But that's where we're headed. That's why Two Thousand Fifty-Two is a crucial year. Two Thousand Fifty-Two will be the year we declare President Goya a One-Term president. Then we will get a REAL commander-in-chief. I don't predict who it'll be. Could be Archer, could be Whitehead, could be somebody else.

"But I tell you this, friends. If we stay on the Goya Path, we're headed down the path of inevitable decline.

"I know some say, 'Why are we even fighting the Caliphates? There's no oil anymore. The whole reason for involving ourselves in that part of the world was for oil, so what is the point now, when our own natural gas fields replaced foreign oil, and then themselves became depleted? What's it all about?'

"I would think we would have learned by now that we should

never ask that question, 'What's it all about?' It's thinking like that that lost us the Vietnam War, the Iraq War, the Afghanistan War, the Pakistan War, the Syrian War, the United Caliphate War No. 1 and part of No. 2, although we're still fighting No. 2.

"But we're losing it, thanks to Goya and his mealy-mouthed, bet-hedging advisors, and the nay sayers in Congress. We should never, ever question the need for a just war. And remember that any war the U.S. of A. fights is a just war. It's our purpose to spread justice and free markets throughout the world, and therefore any war we undertake leads to a more free and enlightened world.

"Those benighted camel jockeys in the Caliphates think we want to squelch their religion. They should be thanking us! If they would welcome us instead of killing us, they would soon find themselves wallowing in prosperity and universal good will. This is why we have to maintain our propaganda outreach to those unenlightened nations, while at the same time keeping the military pressure on, on all three fronts. Instead, we're walking away from two of them. Why?

"Because Goya hates free markets, that's why. He hates the American Way. He doesn't believe we are an Exceptional Country. He said so in his famous Exception to Exceptionalism speech at Columbia University earlier this year. My friends, a man like that should not even be president! No one should be president who does not believe in American Exceptionalism. Goya is the first and only president who has ever taken exception to Exceptionalism. I have no doubt he'll be the last, because the people reject such treasonous viewpoints. He will be resoundingly defeated in '52, if he's not impeached sooner.

"That's my prediction. You heard it here first."

25

My Famous Uncle

The summer wore on. Tom remained in jail, and it seemed like nothing was happening with that case against him. Then Irene said the trial was going to be held in September, and I got a call from the Public Defender, Jade Bessemer, about it. She told me I'd be getting a subpoena, me being the star witness and all.

But Irene didn't have the bail money, so Tom waited it out behind bars.

One night Kareena and I were over at Irene's in an attempt to help her maintain stability in that nerve-wracking time. We ate our dinner wafers and put a movie on the wall vid. As we were settling in on bean bag chairs on the floor, Irene said she was putting Tommy to bed, and she left the room. A few minutes later, she came back, crossed the room to the kitchenette, reached into the fridge and came back with a double-sized bottle of Moscato and three plastic glasses. She poured us each a glass, and I noticed she gave me double what she gave herself and Kareena.

We watched the movie and sipped our wine—I wondered how she could afford it—and whenever my glass got low, she re-filled it. She didn't even ask me. It didn't seem like her. Usually, she was on the borderline of hostile to me.

After a while, I realized the wine must have put me in a stupor, because I jerked awake and realized I'd been sleeping through several scenes. I'd lost the thread of the vid. As I glanced over at Kareena on the bean bag next to me, I saw she was flat out asleep. And then I felt something on my shoulder and I realized Irene was half-asleep, with her head on that shoulder.

"Irene?" I said.

She woke up but seemed pretty wobbly. She wanted to put

her head back on my shoulder.

"Hey," I said. "Maybe we better go. Looks like everyone's asleep."

"No, no, don't go," she said. "Lonely. Lonely."

So it was getting to her, I guess, being alone without Tom. I figured she must be pretty far gone if she was feeling friendly toward me.

I woke Kareena up and we made our excuses and went home.

It was a couple days after this that my mother called me to tell me that my famous uncle had died.

26

Half the Same Genes

My uncle was named Preston Keys. He was a scientist, a clima-
tologist. He studied weather patterns. He was a very important
guy, who was always going to Washington to give reports and go
to conferences. He lived in the D.C. area.

Uncle Preston's wife had died a few years before, so he was
alone when he went. My mother said his housekeeper found him
dead in his study. My mother said she was going to Virginia to
see to the estate matters. She said there might be some things of
Uncle Preston's that would not be of any value to anyone that I
might be able to have, but she wanted to do everything right and
get legal advice.

"Your uncle was a great man," she said. "You may not have
been to college, Jeff, but you come from a good family. And
you're not stupid. You've got half the same genes as your uncle,
remember that."

I said I would remember that.

Uncle Preston was her brother—no relation to my dad.

I could tell my mother was having a hard time holding it
together, even though she tried to sound strong. I could hear her
voice almost crack when she told me I came from a good family.

It wasn't her fault that I don't have a higher education. I have
more education than most people these days. It costs a lot of money
to go to a college and there are no jobs when you get out. There
also aren't as many colleges as there used to be fifty years ago. So
if you find one, you've got to pay. Tom blew a whole wad when he
took his few months of classes at the two-year college in Mansfield.

Anyway, Mom went off to Virginia, and I went there myself
right after that, to attend the funeral, having no idea what was to
result from that trip.

27

In My Uncle's Study

Mom sent me some money so I could attend Uncle Preston's funeral. I took a flight out, and the funeral was held the day after I arrived.

There was a large gathering of mourners, so many that it was held outside the funeral home, in a park across the street. It wasn't the cemetery. They took us in wind-powered hearses to the cemetery afterward, and there was another brief, graveside ceremony there. Some of the mourners were faculty from the university where my uncle was a professor, and some were people from government, and the governor of Virginia was there with his wife. The only relatives were my mother, me, and some relatives of Uncle Preston's deceased wife, my Aunt Zenith. My aunt and uncle had no children.

So my mother, Fleetwood Claymarker, was next of kin and stood to inherit. Uncle Preston had left a will leaving everything to her, except for some bequests to pet charities of his. The will also made a special bequest to me of fifty thousand dollars, with another two hundred fifty to be distributed on my thirty-fifth birthday. So the whole amount was about what you'd make in two years, if you had a decent job, and he'd had this idea that I'd know how to use it more wisely when I was thirty-five than I would now, which was probably true.

Of course, my mom made out like a bandit, and I guessed one day I'd inherit from her, since I had no siblings. This was reassuring, because life is hard. And it gets harder all the time because of the weather and the refugees, and all that.

After we'd had all the mourners over to Uncle Preston's large home in Prince William County, and after they'd left, my mother took me upstairs into my uncle's study. The stairs were carpeted with this thick celadon half-shag and there was a curving polished oak banister leading to the upper floor, so it felt very luxurious going up there. It also smelled pleasant, like putting on a wool sweater.

Inside the study there were book and vid shelves on every wall and desks covered with computers, laptops, tablets, papers, vids, and books made of paper. There were file cabinets with alphabetical markings on them, and they were chock full of files and different kinds of drives. This was only the study, mind you, not the library, which was downstairs and was enormous.

"Look at this pile of stuff," said my mother, her hands waving over it, as her small white-haired head bobbed from side to side to take it all in. She was hunched over from her osteoporosis, so her normal line of vision was aimed at the floor. "I don't know what to do with it. I'm sure it's not worth anything, but I hate to just throw it out. I tried screening some of the videos and mostly they're lectures and recordings of Preston giving talks, or reading research. I'm sure nobody wants it. I wondered if you'd want any of it. You could probably learn a lot watching the videos and reading the manuscripts, Jeff. I was always sorry you couldn't go to college."

Mom knew I was kind of a geeky vid-and-book hound.

"Sure, I'll take it."

"Well, you might want to go through it first to see what looks most interesting. It's not something I'm prepared to do."

"I'll take it all," I said.

28

My Uncle's Vids

My mother financed the transfer of my uncle's papers and vids to my little trailer rental, and the stash took up all the space I had. I still had work through the T-People, and I figured once I got my little inheritance I'd look into getting a bigger place. Maybe by then one of my temp assignments would have turned permanent, I figured. That would be a big help.

In the meantime, I started reading the papers and screening the vids.

I came across one labeled "Deluge of 2025." This subject had always interested me, so I slipped it into my tablet and saw my uncle appear, sitting in a plush-looking burnished leather armchair beneath a standing lamp, smoking a meerschaum. In this vid, he was a man of about forty-five years of age (the date on the wrapper was 2027), with a full head of dark hair, combed straight back, a premature touch of gray about the temples, another gray patch in his mustache. He wore a blue blazer and a tie, tan slacks, and dark blue slippers as he sat with his legs crossed.

I watched intently as he spoke.

"My name is Preston Keys. I'm professor of climatology and biogeochemistry at Vendome Technical Institute in Quantico, Virginia. In this video blog, I will be discussing the effects of the recent Deluge of 2025, including what we can expect going forward.

"From a climatological standpoint, the history of the Deluge is by now well known ..."

29

DNA

Sometime around the end of August, I received my subpoena to Tom's trial. Since I lived in the same park as Irene, I decided to drop in over there to see if she'd gotten one too, or how she thought things were going. I was shocked to find her with all her belongings packed in boxes.

"You," she said when she opened the door to see me. "What do you want?"

I told her why I'd come. I asked what all the boxes were about.

"I'm leaving," she said. "Tommy and I are going north, where it's not so warm."

"What about Tom?"

"What about him?"

While we were involved in this exchange, she was busy packing a suitcase that was lying open on her sofa. Little Tommy was standing nearby, wearing a diaper and a tiny yellow T-shirt, holding a small stuffed bear by the arm, wagging it every now and then. Suddenly, she whirled and assaulted me with her words.

"He killed that girl! Did you know about it? Did you have anything to do with it? You must have known the truth all along!"

She started to cry and sank down on the sofa next to the open suitcase, which was brimming over with clothing. Tommy started to cry too then.

"Irene, what are you saying? You know Tom would never do a horrible thing like that."

"Yes, that's what I know, but what I know doesn't mean anything. They have DNA evidence! They found Tom's DNA in blood swabs from the shack where he took her. His lawyer wants him to plead guilty to avoid the death penalty. So what do you have to say now?"

I was too stunned to say anything at first. At last I stammered something about the subpoena. There must still be a trial, because I was ordered to testify at it.

"That's because he's stubborn to the point of stupidity. He'll go ahead with his trial, and he'll be found guilty, and then they'll put him in the laser chamber."

"Nobody gets the laser chamber anymore. Capital punishment is dead, Irene."

"Oh, no, it's not, not in Ohio!"

It was true; we still had the laser chamber, but it was hardly ever used except for terrorists and the worst serial killers.

"How do you know they have DNA evidence?"

"He told me. His lawyer showed him the lab reports from the Prosecutor. I can't stay living with a murderer, Jeff, so don't try to talk me into staying."

"Aren't you coming to the trial? Are you a witness?"

She wiped her face dry with a towel from the suitcase and seemed to get control of her raging emotions for a second.

"I'm not a witness. Will I go to the trial? Maybe I will. But after that, I'm leaving. I'm tired of living off gruel and wafers and rationed water. Don't you see what's happening to this country? The entire western half of it is a desert. The eastern half is being swallowed up by the ocean. The farm land is gone. There are nothing but refugees everywhere you look. We've got to get to Canada."

"Yeah, everyone wants to get to Canada, but they won't let you in up there. They'll deport you."

"There are ways to get around that."

Now, just as suddenly as she'd calmed down, she burst into tears again.

"Jeff, why did he do it? What did he want from her? Why, why, why did he do it!"

Tom's little trailer had become a madhouse—Irene sobbing, Tommy screaming, Tom's chance at freedom falling apart like rain gutters in a tornado.

I sat down on the sofa, feeling like I'd been elbowed in the solar plexus by an NFL linebacker.

"I can't believe he did it," I said. "It's got to be some kind of mistake. Tom is not that kind of person."

"You can't fight DNA, Jeff."

"I've got to go talk to him," I said, jumping up, even though there was no way the sheriff's people would let me visit Tom until Saturday. "Irene, don't desert Tom now. Wait until after the trial. He could be found not guilty."

"Oh, no, no, you fool, he's going to be found guilty. He's going to be found guilty!"

I left her sobbing in her disorganized living room and walked all the way downtown to stop in at the jail, but they turned me away, like I knew they would. Everything from here on out was going to be like knocking my head against a granite wall. But I just couldn't believe Tom was guilty, and I wouldn't believe it unless I heard it from Tom's own mouth.

And that's why I went to see him on Saturday, during visiting hours.

30

Pajamas

"I didn't, I didn't, I didn't," Tom said over and over from the other side of the glass barrier.

He sat with his head in his hands on the black plastic chair. The light was bright and white on the floor on his side of the barrier, but in my narrow space it was dark. He kept dropping the phone to put his head in his hands, and I kept saying, "Pick up the phone, Tom."

"It's some kind of frame-up," he said, replacing the phone connection.

Tom did not look good. Two and half months of jail had caused his complexion to go pale. His sandy hair was disheveled, as if he had just been sleeping, and it was hard to get him to concentrate. Part of that just-sleeping look came from wearing the orange, jail jump suit, which looked like pajamas, and the sheriff's people always had the prisoners wearing open-toed slippers on their feet, adding to the effect.

Tom looked up at me with bleary, bloodshot eyes.

"You believe me, don't you, Jeff?"

I knew he would ask me this, so I answered quickly.

"Yes, I believe you, Tom. But how did they get DNA?"

"It's some kind of plant. It's some kind of set-up."

"But how, Tom? Were you ever in that shack?"

"Never! I was with you, don't you remember?"

"Yeah, sure I remember."

The funny thing was, I really did believe Tom. At least, something in my heart believed him. My instinct believed him. I knew him, and I could not believe he had killed that girl, but my rational mind said there was no way an innocent man would have his DNA mixed up with the girl's blood samples taken from the crime scene.

"Irene said your attorney wanted you to plead guilty, Tom."

"Irene! When did you talk to her?"

I told him about my visit—and I couldn't help it; I also told him about the suitcases.

"She's leaving me," Tom moaned. "She's given up on me. My lawyer's given up on me. Everybody's given up on me. But I'm not giving up! I'm not going to plead guilty, Jeff, because I did not do this crime. They're going to have to get twelve people to say I did it, and they're going to have to drag me to prison, because I'm innocent. Jeff, you can't let Irene take little Tommy away from me. Don't let her do that!"

I changed the subject back to the original one, because the subject of Irene and Tommy was hopeless, as far as I could see. "Tom, you said someone was setting you up."

"What else can it be? It has to be a set-up."

"But why, Tom? Who do you think did it?"

"The people who killed her. That creep who stole my water card. Who else?"

"But we never saw that guy before."

"I know that."

"So why does he want to set you up? He doesn't know you from Adam."

"How do I know? So he doesn't take the rap. What else? What other reason could there be? How the hell do I know anyway!"

I went away from this interview racking my brains. There is really only one reason why somebody frames someone else for a crime, I thought. And that is to throw suspicion off themselves. But that kid that held us up was nobody we knew. And we only encountered him for a minute or two, just long enough for him to steal the water card. And no matter how I turned it over in my mind, there was no way I could get from that encounter to the fact of Tom's DNA on blood samples from a shack where the girl had been killed.

It depressed me to think it, because it meant that I couldn't trust my own perceptions. It looked like I was no good judge of character, because it had to be that Tom really did kill that girl, Hypnotica.

31

There Would Always Be Enough War

From that day until the trial, I spent most of my time between my temp job and watching the vids of my uncle. I saw Kareena, of course, but I was worried that we might be getting in a little too deep. I just couldn't see pairing off with Kareena on a permanent basis. And then I remembered what Irene had said, how I was a lizard and all that, and how guys like me thought we were God's gift to women. And I wondered if it was true, that I was really a selfish egomaniac who overvalued himself.

I thought maybe I would have a talk with Kareena, because she might have her own take on things that might prove to be enlightening. But I wasn't going to make any moves at all until after the trial.

So the only thing I focused on was the vids from Uncle Preston.

I was amazed at the stuff he had recorded. It was like having your own philosopher explaining everything to you. It was like having the father you wished you had, telling you how everything could be looked at. I mean, most of it was professional stuff, but some of it was just him talking into the camera, commenting on life. Like I found one labeled, "The Ordinariness of Life," and this is what my uncle was saying in that one:

"When you're young, you're full of dreams, and you live from day to day, thinking one day this will happen or that will happen, and it is always something wonderful and thrilling. Then, as you age, you find that life continues from one day to the next, and nothing much ever happens—at least, nothing very extraordinary. And if you don't make something happen, there is no fate that makes it happen. Instead, fate waits for you to take the lead. It's up to you to set something in motion.

"Most of us make a career choice, and that choice is what opens up the path of the future to us. Our social lives follow from whatever choices we've made—not from a lightning bolt from the sky, but from ordinary choices and the chances that arise from them. If we stay in our homes all the time and never go out, nothing much happens to us. If we venture forth, things begin to happen, but it is never something that could not have followed from the choices we made.

"Very often, we find ourselves disappointed in some way. Life is full of setbacks, dreams that don't materialize, compromises. We learn to live with these compromises, and we ultimately come to be defined by them.

"And then, one day, we're middle-aged. We are well-defined, and our definition limits us. The world is no longer open to us as it was when we were young and full of dreams. Yet, in another way, it's more open than it was, because we have a reality which we didn't have in our youth, and we carry out this path of self-definition, which is perhaps not glamorous or extraordinary, yet which has a certain fulfillment of its own."

This talk scared me a little bit, because it sounded like I could expect only a kind of mediocrity as I got older, but I had to admit that I was already seeing the truth of it, because here I was, twenty-six years old and no wild and wonderful fate had befallen me. In fact, I was living out exactly what I guess I might have expected all along, given the state of the economy. I mean, I was working at temp jobs, doing manual labor, and not getting ahead in life much.

I used to think, when I was young, that I would do some exciting, brilliant thing that would set me apart, and would show everyone that I didn't need a higher education, because I had natural talents. I used to sing a little, and I thought maybe I would become a famous singer, but that never happened. Then I thought I might write a book, because I like to write, but I sent some stories out to online magazines and nobody was interested.

And as I would read the stories over, I could see why no one was interested, because I really didn't have much to say, and everything I wrote sounded just like I sound when I talk, which is not anything super special, I guess.

So I could see what my uncle meant about how life just kept on going from day to day in a very unremarkable fashion. And also how it shaped itself around whatever you had set in motion. And I could see that I hadn't set very much in motion, but then, how could I, in the world we lived in? The economy was depressed. There were no jobs, even for well-educated people. Everyone just kept waiting for the next hurricane, the next tornado, the next rainstorm, the next pollution alert. And the refugees kept coming from the West, and then going North, like Irene planned to do.

I listened to Wesley Wright talk about how bad it was overseas. Drought and famine and dislocated populations were everywhere, and they were worse than they were here. It seemed like the only reason we had people left in the USA was that it was so much worse everyplace else, that they kept coming here to escape whatever it was someplace else tsunamis, floods, earthquakes, heat waves, and, mostly, lack of food.

There were plenty of guys—even guys as old as me—who gave up and joined the army, just to have some kind of future. It seemed there would always be enough war.

32

I Am Cross-Examined

Tom's trial took place. They wouldn't let me in the courtroom until it was my turn to testify, because, they said, they didn't want me to be influenced by what other witnesses said. So the only part of it I saw was when I went in to testify.

The courtroom was packed with spectators. There wasn't a free space in the room. The spectators were seated on long wooden benches like church pews, and the jury was on the side of the room in these blue cushioned chairs near to where the witness stand was, next to the judge's bench. A bailiff came out into the lobby to summon me and I went into this scene and strode up to the witness stand and took the oath to tell the truth. Then I sat down in this little chair that was all by itself in a wooden box with a door on it. Everybody was looking at me: the prosecutor at his table, with a policeman next to him, the Public Defender at her table, with Tom sitting next to her, the people in the pews, the jury, the judge, the bailiffs—everybody.

On the far yellow wall was a clock with the hands at three. I'd been sitting out in the hallway all day, and it was a relief to finally get to come in and get my story over with.

Tom's attorney questioned me first. She was wearing a black dress, or suit, with white shirt and collar. Tom was sitting next to her, wearing his street clothes. He told me, when I went to visit him in jail, that he would not be wearing the orange jump suit, because that would make the jury see him in a bad light, and that wasn't fair. So he had on a patterned sports shirt and brown Levis and black dress shoes. It was strange to see him sitting there, and even stranger to think that he could have murdered someone. I knew they had DNA evidence, but somehow I continued to find the accusation unbelievable. Tom had never been a violent person. But as I looked around at the crowd, and at the jury, I realized that no one there knew him as I did. Not one of them had any reason to believe that Tom was not capable of murdering someone.

And I thought, all of us are subject to any belief that someone adopts about us, if they don't know us. And most of the time, it doesn't matter what people think. But in this case, everything depended on it. And none of these people knew Tom.

Ms. Bessemer asked me a few questions about myself. What was my name? What did I do for a living? Did I know the Defendant? How long had I known him? How would I describe my relationship with him? Did I know his family, and so on. Then she said she was calling my attention to April 4th of this year. Did I recall where I was? Did I remember what I was doing? Who was I with? What happened?

So I told the whole story of what happened that night we went to Tent City, and how we'd been held up by some kid with an Encroachers black T-shirt and tattoos on his arms, and how he'd stolen Tom's water card. I explained why we didn't call the police. I explained that afterward, I didn't remember very well, except that I went home and Tom went home, that I knew he went home because he only lived a few doors down from me. I explained that I didn't know the boy who'd stolen the water card, had never seen him before. She asked if I believed Tom would kill someone, and I said I found it impossible to believe, because I knew him, and he was not a violent person. She asked me if I believed he had done it, and I said no, I personally did not believe it.

Then the prosecutor got up to ask me some questions. He was about thirty-five years old, a minority white guy, wearing a navy blue suit and a red tie and white shirt. Medium height, average build.

"You're Tom's best friend, aren't you?" he said.

"Yes, I would say so."

"You'd do anything for him, wouldn't you?"

"To help him, you mean?"

"Yes."

"Sure."

"Even lie for him?"

"No, I wouldn't lie."

"But after this alleged theft of Tom's water card, you say you think you went home and Tom went home?"

"Yes, we didn't go anywhere else after that."

"So you don't know where he went afterward, do you?"

"Other than he went home, because I saw him go home."

"But after that, he could have gone out somewhere, and you wouldn't have known about it, would you?"

"I guess not."

"You weren't with him after that?"

"Not that night, no."

"Some other night?"

"Well, I saw him almost every day. But soon after that, the police came to interview him."

"You didn't go with him to the shack out by the fairgrounds?"

"No."

"He went there without you?"

"I have no knowledge of him going to any shack."

"I suppose you have no knowledge of him meeting a woman and asking her for sex?"

"No, I don't. He has a wife and child."

"You have no knowledge of him murdering a woman and cutting up her body?"

"No."

"You never saw him acting strangely or suspiciously, I suppose?"

"No."

"You think he did not kill Hypnotica Christiansen?"

"I can't believe he did that."

"Even though his DNA was found on blood samples in the shack where she was killed?"

"I heard about that."

"You don't think his DNA proves his guilt?"

"I don't know what it proves."

Jade Bessemer was objecting at this point, but I was excused shortly afterward. And so I left the courtroom, feeling sick.

33

Palm Trees in Alaska

By the time Tom was convicted and sent off to prison, his rented trailer down at Lot 525 was empty, abandoned, windows blank, a red "For Rent" sign in the door. Irene and Tommy were gone: where to, I couldn't say. She had said Canada, so I assumed she'd gone up there, but I knew that wouldn't be too easy, because the Canadians were cracking down on illegal immigration. Their population had tripled since the Deluge of '25.

All populations were moving north, and had been all my life. The area where I lived hadn't had near the population it did after the Deluge, which occurred the year I was born, and those were people from the South and the West. California and Arizona were emptying out, had been for decades. There was no water there, and temperatures in summertime were a hundred twenty, a hundred twenty-five—temperatures that would kill you, if you weren't properly air-conditioned. And Mexico, forget it. We were practically at war with Mexico and all the people from other countries trying to come up here from down there.

Alaska was booming, despite some water problems of its own, but there was a lot of resentment among the Alaskans at all the newcomers too. In fact, there was a secession movement up there, but everybody knew that would get nowhere, because the federal government, weak as it was, still had all the soldiers and military equipment and could easily put down any rebellion.

Some people said that Alaska had only recently put in some paved roads up there so you didn't have to get everyplace by pontoon plane.

Also, somebody had planted palm trees in Alaska. I saw a picture of it in the vids.

So now Irene was moving north along with the masses ...

One night Kareena called me on my handheld. I thought she would ask me over. That was usually why she called. But tonight she said, "Do you think we're seeing too much of each other?"

"I don't know. What do you think?"

"I think we are."

This caught me by surprise. I had the feeling that she was saying this because of the way I'd responded. I mean, like, if I'd said, "We are absolutely not seeing too much of each other," she would have asked me over, but instead, because of my non-committal response, she'd decided to slap me down.

"You think we are," I repeated back at her.

"Yeah, well. I think a relationship either gets better or it gets worse. And ours hasn't been getting better."

"In what sense? We have good sex."

"Good is good is good, but it has to get better, or it gets worse—the relationship, I mean."

"Are you talking about commitment?" I said. "Because if you're talking about commitment, you should say so."

"Well, there is more to a relationship than just sex," she said.

So she was talking about commitment, I realized. But I didn't want to have that conversation right then.

"Well, if you don't think things are getting better," I said, "and if you think we're seeing too much of each other, then I guess we should see less of each other. Is that what you're saying?"

"Maybe I am," she said.

"Well, all right. So when do you want me to call you again?"

"Maybe not for a while."

"Oh, really?"

"I think I should tell you that I have another friend," she said.

That came down hard, like a mallet. I should have guessed it sooner.

"So I suppose you don't want me to come over tonight?"

"Not tonight, no."

"Maybe not for a while?"

"That's up to you."

"Is it, Kareena? Is it up to me?"

"Oh, I don't know. I don't think I want to talk anymore."

"All right, then."

"All right, then."

And she cut the connection.

That was a hard blow. It seemed like my life was tumbling down on my shoelaces, what with Tom going to prison, and Irene leaving town, and now this. I thought, I'm a miserable failure—a pathetic excuse for a man—a worthless parasite without a permanent job—a man with no future—and now a man with no friends.

Outside, it got really black, and the lightning bolts started firing up.

34

Serendipity

There's this thing called Serendipity. I think it's like some kind of weird, surprising good luck. I saw a vid once where somebody was saying the idea came from a Persian fairy tale, but I don't know anything about that.

Anyway, I had some kind of Serendipity less than a month after Tom went to prison. It happened while I was going through my uncle's vids. I came upon this stash of them that were marked 'HIGHLY CONFIDENTIAL: NATIONAL SECURITY.' That was what was on the small white box that held the flash drives. I took the drives out and started playing them. I was quite surprised by what I learned.

Here was my uncle, now aged, I guess, about sixty. He had a short white beard and was addressing the viewer directly, as he did in most of the vids. This is what he was saying:

"I have had occasion to work closely with figures in highly confidential positions in the national government. This has included work for the various bureaucracies associated with our National Security network. In the process of doing this work, I have gained access to a great deal of information—much of it only in the form of brief glimpses at file names. Nevertheless, I was surprised and disturbed by much of what I saw. The government has instruction vids for all sorts of clandestine operations and dirty tricks. For example, networks for controlling information in the major media outlets, techniques for transferring DNA from a stolen water card in order to frame someone for a crime,

methods ..."

I lurched forward in my chair and a cold chill went up and down my body as a shock went straight to my heart.

Techniques for transferring DNA from a stolen water card in order to frame someone for a crime … !

I couldn't believe what I was hearing. It was like that movie, Hamlet, where the ghost comes back from hell to warn his son who murdered him.

Techniques for transferring DNA from a stolen water card in order to frame someone for a crime … !

I felt faint. I played back the section on the recording to make sure I'd heard right. When I was sure I'd heard right, I knew what had happened to Tom. It would have been incredible, except that I remembered how the boy, Milteer, had known my name. If he had been simply a lowlife criminal on the street, there's no way he would have known my name.

Tom had been the victim of an elaborate plot. But why? There was no reason to select him.

I played out more of the vid, but Uncle Preston did not refer again to DNA or stolen water cards. It had been a passing remark, but one coming from an inside source, someone who had been in the offices where National Security was implemented.

There was some kind of incredible story underneath all this, but I didn't have a clue what it was. I only knew that Tom really was innocent, and that I had to try to find a way to get this information to someone who could help him.

I went out to the nearest station to get some water, because I felt really dry. I stood there in line with my card and finally got a twenty-ounce bottle and drank half of it right there. Then I went out walking in the rain.

I developed a serious headache by the time I returned to my trailer. I stripped off my clothes and dried myself off with a towel and watched some entertainment vid until it was time to go to bed. Then I lay awake most of the night, with the sides of my head throbbing in terrible pain.

35

The Nearest Gruel Kitchen

I didn't have any work that week, so the next day I walked down to the square to see if I could talk to the Public Defender, Jade Bessemer, about my discovery that Tom had probably been set up. I took an umbrella, and used it, but the rain had slowed to a continuous drizzle and wasn't a problem. The sky was the color of dark smoke. Fall had arrived and the wind was nippy. My little dun-colored baseball jacket was not warm enough to keep away the shivers.

I stepped through the heavy glass door and entered the lobby from the street, and then entered the Public Defender's office through a heavy wooden door leading to a pair of windows where the secretaries sat. There were two secretaries, and neither one of them looked up when I came in. I was the only other person present, and I stood at the window until one of the secretaries finally looked up. She was a heavyset woman with gray hair and rectangular eyeglasses.

"Can I help you?"

"I'd like to see Ms. Bessemer, please."

"Do you have an appointment?"

"No. If I need to make one, I can—"

"Are you charged with a crime?"

"Me? No, I'm a friend of—"

"So you're not a client of ours?"

"A client? No, I'm—"

"Well, she won't see you if you're not a client. And if you're not charged with a crime, we can't do an eligibility intake. What are you here for?"

"I'm here about Tom Glendinning. I was a witness in that case. I've found—"

"He was sentenced to life in prison. That case is over."

"Yes, I know that. The thing is—"

"Well, we're not involved with it now."

The woman returned her gaze to her screen, but then she thought of something and turned her eyes back on me.

"Are you looking for Mr. Glendinning's appellate attorney, maybe?"

"His—?"

"Are you here about his appeal?"

"His appeal?"

I hadn't known there was an appeal.

"Well, what if I am?" I said, hoping to get some information instead of a roadblock.

"That's being handled in Columbus. By the Ohio Public Defender."

"Well. Do you have their number?"

She brought it up on her screen and wrote it down on a card for me.

"Thanks," I said, slipping the card into my shirt pocket.

The secretary's attention had returned to her screen, so I figured I was dismissed. I went back into the lobby. It was a nice-looking lobby, with a green marble table affixed to the floor by a concrete stand and a pot of fake, plastic flowers in a heavy gilt flowerpot nearby.

I still felt a little weak from last night's headache, but the pain was mostly gone now. I felt light-headed, as if I'd been in a boxing match and had taken some hard punches to the temples and been made dizzy. I felt really hungry, too, the way you feel when you want something more than standard gruel or government-issue wafers, but I didn't have enough money on me for anything else, so I walked down the street to the nearest gruel kitchen and had breakfast.

36

Algae Blooms in Lake Erie

The gruel kitchens were not-for-profit restaurants where you could trade water credits for food. The food was not the greatest, but it was cheap, and you could get things other than gruel. That term came into use as a put-down, because the menus were limited.

The place I went to was called Sal's and it was on the corner of Liberty and Walnut Streets, a block away from the courthouse and the public defender's office. It had a counter where I sat on a blue rotating stool and ordered a small bowl of diced potatoes seasoned like hash browns, and a cup of tea for five water credits and eight dollars cash.

Feeling a lot better after I ate, I headed across the street to the public library, which had just opened its doors. The library was a good place to make a cell call from, because you could usually find a quiet table in a corner somewhere, and there'd be no street noise.

I punched in the number of the Ohio Public Defender on my cell and reached another secretary.

"Ohio Public Defender."

"Hello, my name is Jeffrey Claymarker. You're doing an appeal for Tom Glendinning, who was convicted of murder. I was one of the witnesses at the trial. I came across some information that shows Tom wasn't guilty."

I felt like a telemarketer, talking fast to get the information in before the hang-up.

"I can't help you with that. You need to talk to an attorney."

"Yes, I know. That's why I'm calling. I was referred to you by the Public Defender here in Wayne County. They told me to call you and—"

"Just a minute."

I heard the secretary asking somebody a question, but her voice was faint. Then she came back on the line.

"The attorney handling that case is Mr. Hockles. I don't know if he's in, but I can transfer you to his extension. Do you want me to do that?"

"Yes, please."

After a couple of rings, a man's voice came on the line and announced himself so quickly that I would not have been able to make out the name if I hadn't already been given it.

"Mr. Hockles," I said, speaking quickly, "I'm calling about Tom Glendinning. I'm Jeffrey Claymarker. I was a witness at Tom's trial."

"Claymarker! That's you?"

"Yes."

"I know who you are. I've read your testimony. I've got the transcript right here. What can I do for you?"

I explained that I'd learned that the government had secret programs for framing people for crimes using transferred DNA from stolen water cards, and that my uncle did contract work for the National Security Agency, and—I was interrupted.

"What! Where is this coming from? Who told you this?"

"My uncle. That is, my uncle is dead, but he made these recordings before he died, and—"

"Your uncle! Who the hell is he?"

"His name was Preston Keys. He was a scientist. He—"

"Preston Keys? I know that name. Or maybe I don't. You say he's dead now? When did he die?"

I gave him the story.

"You say he had information about using water cards to extract DNA?"

"To frame people for crimes they didn't commit. Specifically for that purpose."

"That's incredible."

74

"Mr. Hockles, I have the vids. I have a vid of my uncle describing this program in the NSA. I can play it for you."

"It won't do any good," he said.

"It won't?" I was taken aback.

"No. Listen. Does it have anything to do with Tom Glendinning? Specifically?"

"No. But there has to be a connection, because they knew my name. They knew me; that kid that stole the card knew me, and I never saw him before. He can't be simply a street thug. He has to be someone with some kind of surveillance information, someone—"

"Listen. Listen to me."

I could hear in the attorney's voice a number of subtle emotions, attitudes: doubt, impatience, yet some openness, a cautious half-open door.

"That's new information," he said. "I'm doing an appeal. I can't use new information in an appeal. I can only use the transcript. That's all they'll look at. You understand?"

That was clear enough, and my heart fell. I said I guessed I understood.

But then he went on:

"New information is something we use in what we call a collateral attack. A post-conviction petition. It's a different process. What you've got there is a little bit of information. It's interesting, but more information would be better. You need an investigator or something. Look, tell you what I'll do. Send me a copy of the vid you're talking about. Don't send the original, just a copy. I'll look at it. I can't promise I'll do anything with it."

I said I would make a copy and send it to him. He gave me the address and the email. Then he said:

"If you had an investigator—someone who could tie it up better—get some more information about the water card thing. That's what you'd need. I can't promise anything. But I'll look at what you have, OK?"

"OK."

So I made a copy of my uncle's vid and emailed it to Hockles. I waited for a week, expecting he might call me back to say how blown away he was by what he saw, but there was no call.

Before the week was up, though, I got another temporary employment assignment, this time inputting information about algae blooms in Lake Erie for a research firm. I felt like I was moving up in the world.

37

Cancer & Chaos

I started watching this pretty glue channel called Historia. They had these heavy-duty intellectuals on there, talking about all kinds of stuff. There was this one named Professor Brandon Cognovit, who had written a history of the Cold War in the mid-twentieth century.

"The fear of Communism in the West," said Professor Cognovit, "which was used so effectively to control the hearts, minds, and actions of the masses, was a displaced fear of cancer and chaos. In fact, you do not have to look very deeply into the media of this time to see that the Communist threat was often overtly described as a 'cancer', and was allegorized in science fiction films as a creeping threat that surrounded you and threatened to overwhelm you by infecting the bodies and minds of those previously friendly. This is analogous to the cells of the body, which are gradually overtaken by cancer cells and are converted from functioning, 'friendly' cells to hostile, life-threatening cells. The fear of cancer was quite powerful at this time. Whereas today many types of cancer are curable, and most are treatable with significant positive effects, in the mid-twentieth century, it was a fearsome malady, seen as the one disease with No Cure. So we should never underestimate the power of this metaphor of a 'cancer' in Cold War anti-Communist ideology.

"The fear of chaos, on the other hand, is seen in the many references, particularly in the very early days of the Red Scare, to the godlessness of Communism. Marx's famous line that religion was the 'opium of the people' became well-known in the West, as did the hostile policy toward religion taken up by the Soviet regime. And the West, I would say particularly America, was very susceptible to the argument that Communism was a form of mind-control that sought to take away your religion, leaving you to the cold, empty, perhaps brutally scientific realm of a universe without a loving heavenly Father overseeing and making sense of it all. Fear of cancer and fear of chaos are very strong psychological motivators, and they were used, quite adroitly, in my opinion, to garner support for policies that were, in some cases, of rather dubious merit, such as the McCarthy hearings, the Viet Nam war, or the arms-for-hostages deal in the Reagan administration to support anti-Communist proxy wars in Central America, and so on. Eventually, as these policies dragged on and became more and more transparent, people began to see through them, and there was a growing opposition, and eventually the policies collapsed. But these were ad hoc collapses, controllable collapses, from the point of view of Cold War anti-Communist ideology.

"What then developed, during the final phase of the Cold War, under Reagan, was a kind of reification of the old ideas, primarily that we were not crazy to oppose cancer and chaos, that it was, in fact, cancer and chaos that we had all along opposed — the idea of the Evil Empire. Here, again, was an idea reminiscent of science fiction films that had been popular in the seventies — Star Wars donating the idea. Now that we were not fighting a shooting war on any large scale, such cartoonish concepts were not subject to the skepticism that arises when one confronts war on a daily basis and is forced to ask whether the threat is real or not, and worth opposing with large expenditures of blood and treasure. And then, when the Soviet Union fell of its own weight, we found ourselves with a pattern of thinking which had no longer any external object on which to drape it, until it was later possible to fashion a Terrorist Threat, which could take the place of the former Communist Threat and could become a kind of new, potential Evil Empire of its own. The genius of this new incarnation of the idea was that it lacked a corresponding nation state, and was all the more fearsome by being so amorphous. It now lacked any necessary finiteness and could be successfully used to spawn a psychology of endless war, capable of fueling the economic engines of the military industrial complex of which President Eisenhower warned in the early days of the Cold War."

That guy could sure talk.

Whenever I watched Historia, I kept my dictionary vid handy and had it online for immediate real-time definitions.

38

The NFDA

I was only five years old when the laws were passed creating federally subsidized genetically-engineered food substitutes—which led to the wafers and the gruel kitchens that provide the greater part of my own nourishment to this day. I remember the brouhaha in the media about it, but not very much of what it all meant then. To get a handle on it, I had to watch vids about it since then.

There was a good one on the Historia channel, which I downloaded. The expert doing the talking was Dr. Fernanda Stake, who was described as a professor of political science at the University of Michigan. Here are some of the highlights of Dr. Stake's analysis:

"In the wake of the Deluge of 2025, and a number of other climate-related events preceding it, the viability of agriculture and the various food production industries in the United States was threatened, largely from pressures from governments overseas, whose demands for imports at certain levels put our farmers in increasingly fierce competition with those in other countries for world markets. Floods, droughts, losses of livestock due to summer heat waves and large fish migrations and die-offs led to lower yields in most food production categories. The U.S. Congress asserted that interstate commerce was affected, and the National Food Distribution Act of 2026 was the result.

"The NFDA authorized the Department of Agriculture essentially to subsidize large scale production of genetically engineered, low cost foods or food substitutes, which today are known as 'government wafers' and the gruel that is sold in so-called 'gruel kitchens.' The wafers are not, in fact, 'government' created, but rather are mass-produced by private companies with heavy government subsidy, with the goal of feeding as many people as possible with minimal food resource output. This, in turn, permits the agricultural and food production industries, such as the cattle, dairy, and seafood industries to maintain advantageous positions with respect to world markets.

"Many conservatives and libertarians were quite angry about the Act's passage, claiming that it was one more example of a policy engendered in the United Nations, good for the world, but bad for the U.S. However, the Act was passed largely at the instigation and through the lobbying efforts of the huge agricultural industry, as well as the cattle ranchers and the fisheries.

"Political scientist Dominic Rafeld of the University of Texas, Austin, argues that, although the Act was passed under the authority of the Commerce Clause, it was, in fact, foreign policy considerations that drove Congress and the President to support it."

The montage of clips of people eating wafers, of beef steers in a pasture, of fishing boats, and so on, now gave way to an interior shot of Dr. Stake interviewing Dr. Rafeld in a book-lined study. Dr. Stake was an attractive, dark-haired woman in her thirties, and Dr. Rafeld was maybe fifty, bespectacled, with frizzy graying hair and a mustache.

Dr. Rafeld speaks:

"What the President and Congress were worried about was war. We had enough food to eat in the United States, but Africa and Asia were starving. When populations starve, war tends to break out. Guerrilla groups form. They attack governments. They try to take over. Governments attack other governments. There were no less than five wars already heating up—no pun intended—in Africa and Asia when President Lawson signed the bill."

A voiceover from Dr. Stake addresses the audience in mid-interview:

"But a law creating 'government wafers' and gruel kitchens

could not be passed for foreign policy reasons."

Dr. Rafeld resumes:

"Congress can't pass a law to prevent wars under its war powers. It can only declare war. If it's going to pass domestic legislation, it has to do it under the Commerce Clause. Of course, it was largely a commercial concern that was being addressed, but it had consequences that went far beyond interstate commerce."

So that was pretty interesting. They went on to explain that it was an equalizing measure. Political opposition to the Act died down after a few years, because the wafers and gruel were economical, and the worldwide economic downturn meant people were looking for cheap sources of food.

I was too young to remember any of the political stuff. I've eaten the wafers and the gruel as long as I can remember, with some real food on the side when I can afford it. I guess the fact that we use the term 'real food' is a holdover from people who remember things the way they were before the NFDA.

I just think it's ironic that I eat the stuff I eat to keep some guerrilla in Africa or Asia from carrying out a coup.

39

Unturned Stone

September turned into October, 2051. Temperatures hovered in the mid-eighties. The leaves were slow to turn color.

Early in October, I received a paper letter from Anthony Hockles of the Ohio Public Defender's office in Columbus. It was not a long letter, but I read it with high interest. Here is what it said:

Dear Mr. Claymarker:

I have reviewed the video recordings you sent me. I find them interesting. However, I don't find enough in them that, in my opinion, would cause the court to re-open Tom Glendinning's case. In fact, you have nothing but a bare statement from your uncle that a program exists that can be used to make criminal use of DNA from water cards. The existence of a stolen water card and DNA evidence in Tom's case makes this statement very interesting, but does not, in itself, prove Tom's innocence.

If you wanted to follow this up, you'd probably need to hire an investigator, and that could cost a lot of money. It might not lead anywhere.

I'm sorry that I will not be pursuing the matter further.

Best wishes,

Anthony Hockles

So that was that. A recap of the reluctance Hockles had originally expressed over the phone. What a let-down.

I wrote to Tom about all this, and sent him a copy of Hockles' letter, but after a while, Tom stopped writing back. When he did write, he would always ask about Irene and Tommy. I had nothing to tell him about them, because I had no idea where they had gone, or what had become of them.

I think it was the second week in October when my mother told me to come over to her house in Rittman because she had a check for me. When I got there, she presented me with the check from my Uncle Preston's estate for fifty thousand dollars. It was not a fortune, but it was more than I'd ever had.

The first thing I did was to rent a house in a better neighborhood in Wooster, although I remained close to the downtown area. My rented trailer had been getting cramped anyway and I wanted to get out of there. The new house had two bedrooms and a small back yard. It had a wind/electric vehicle garage, but I had no vehicle. I rode a bicycle to get around, or else I walked or took the electric city bus.

I had some refugees from Alabama living next to me, people whose hometown was now under water. It was a black family, mother, father, four kids and a grandmother. I felt sorry for them, and all the other refugees. Their house didn't look any bigger than mine, so it must have been pretty cramped.

One day, I don't know why, I found myself checking the vids for business listings, under the heading "Private Investigators." Some had simply names of investigators, and some had company names. I found one called 'Unturned Stone.' I liked that name, so I called up and made an appointment.

The next day I went down and plunked down five thousand dollars as a deposit to hire an investigator named Cyril Shaker, to see what he could dig up that might help Tom.

40

A Sad, Sad Wreck

This Cyril Shaker didn't look like the private investigators you see in the vids. He was a compact man, maybe fifty or fifty-five years old, with graying dark hair, cut very short and balding in the back. He wore wire-rim glasses and a gray suit with his red tie pulled loose. He looked more like a tired professor than a private eye, not that I've seen a lot of professors personally. I see them mostly in the vids.

His office was downtown, on the square, above a pawn shop and a one-hour EV/WV lot.

When I told him about Tom, and then about the vid from my uncle where he talked about the secret government programs, Shaker got this melancholy look in his eye. He asked if I had a copy of my uncle's vid, and I did. I gave it to him, and he said he'd look at it later, if I hired him.

"Do you think you can help me?" I said. "I mean, do you think there's anything we can do for Tom?"

He looked very sad, and his indigo eyes became haunted.

"There's a great deal we can do," he said. "Whether it helps your friend or not, we can't possibly know until we do it."

So that was when we talked money, and I ended up paying him the five thousand dollars. I signed a contract, but we specifically agreed that I would have to authorize more funds before I was obligated to pay more. I didn't know how high I was willing to go with this, but I thought that splurging a little in an attempt to get justice was not the worst way to spend part of a windfall. What else would I do with it? Go to college at age twenty-six in an economy with no jobs—an economy with a thirty percent unemployment rate?

"I have to know how to reach you at all times—phone, skype,

email, street address, wearable tech codes," he said.

I gave him all the numbers and addresses.

"Now we're a team," he said, appearing more haunted than ever, as if I'd condemned him to some kind of Greek afterlife. "Everything I communicate to you, you must make a record of it. Do you understand?"

"Yes."

"And you'll do it?"

"Yes, sure. I'm good at keeping records. Don't judge me by my job."

He cast a sharp glance at me, and made a peculiar remark.

"Oh, this economy is a sad, sad wreck," he said.

41

Geometry Out of the Question

The algae bloom assignment lasted only a couple of weeks, but then I got lucky and landed a permanent job doing stock work for a supermarket. And that was where I met Nola Sheeka.

The job came about through the temp agency, but I was only temp for a week when the store manager, Mr. Barrington, asked me if I'd like to go permanent. I said I would.

Mr. Barrington was about sixty years old, about five feet nine in height (a couple inches shorter than me), about one-eighty in weight, mustache, glasses, graying brown hair parted in the middle, pale skin. He seemed like someone who had more going on in his head than he let on, by which I mean he seemed too intelligent for his job. But you can never really be sure about things like that unless you know someone well.

Anyway, Mr. Barrington made a pleasant boss. He would be my direct supervisor in the liquor section. He took me around and introduced me to some of the other supervisory staff, but it was a big supermarket and there were plenty of workers I didn't meet and wouldn't meet until I'd been on the job for a while.

One of them that I did meet early on was a strange character who also worked in stock, and he thought he was doing me a favor by befriending me and behaving in a familiar manner, as if we'd known each other for years. His name was Jeog Hibben. I know it was spelled that way because it was written on his shirt, stitched in above the shirt pocket. He told me that he had that done himself; it wasn't something that came with the job. His name was pronounced to rhyme with 'rogue'. I tried to be nice to him, but I soon noticed that he was not right in the head. Some kind of eager neediness or insecurity made him overstep ordinary manners and customary distances.

We were introduced by Mr. Barrington.

"This is Jeog," he said. "He stocks the produce section. If you have any questions, and I'm not around, you can ask him."

After Mr. Barrington left us, Jeog started right in asking all sorts of questions, some of which seemed peculiar. The peculiar ones were mixed in with the more straightforward ones, like this:

"Where do you live?"

"How long have you been out of work?"

"Do you have any pets?"

"What's your favorite kind of music?"

"Are you married?"

"Do you have a girlfriend?"

"Do you like people who like parakeets?"

"Do you think my shoulders look pointed?"

"How many bottles do you think you can stack before they fall over?"

"I'm in love. Are you in love?"

To be honest, I can't remember the exact order of these questions, because he would pepper me with them, one after the other, but I remember these.

I remember that question about the pointed shoulders because I did think his shoulders looked pointed, but I shook my head and said, "Not particularly," when he asked about it.

He had a kind of tall, oblong body generally that looked as if the various components—head, shoulders, torso, legs, hands— had been slightly deformed in the womb, or coming out of it. I don't mean that he looked monstrous in any way, only that, for example, the head was not quite round and not quite oval—not quite any regular shape, but rather sort of round here, sort of triangular there, as if it couldn't make up its mind; and the same was true with the other parts. Geometry was out of the question, if that makes any sense.

This is not to try to be ungenerous to Jeog or anything. I mean, I'm no great prize in the looks department myself.

The girl he was in love with was Nola. She worked in the store. Had I met her yet? No, I hadn't.

"You'll meet her soon," he said. "You'll probably fall in love with her. She just broke up with her boyfriend. I'm going to get her to marry me."

I thought, given the kind of flaky way Jeog talked, that this Nola would probably prove to be as strange as he was. I mean, he didn't inspire me with confidence in his judgment, so I figured she was probably a loser.

All of that changed when I did meet her.

42

Nola

It was my third or fourth day on the job. I had passed through the stock room swinging doors, into the large portion of our buildings behind the store, and was headed to the break room when I practically collided with Jeog Hibben. He was beside himself with excitement.

"She's back there," he said breathlessly. "The woman I love. The woman of my dreams. Are you going to the break room?"

"I was," I said.

"Then you're going to see her. Remember: she's mine. I'm going to marry her. So don't fall in love with her."

He had grabbed my shirt with both hands when he said this. It was startling, to say the least. I raised my hands to my sides and waved them non-committally and shook my head, as if to say, "Hey, I'm cool."

"I've got to go back out," he said, evidently tortured by the thought of my being in the break room with his fiancée while he himself was duty-bound to return to work. Then he disappeared through the swinging, port-holed doors.

I smoothed the crumpled wrinkles on my shirt front and proceeded down the path, past the stacks of metal shelving and the just-arrived shipments of grocery items, about sixty percent of which consisted of wafer or gruel variations. (I heard later that they gave the produce section to Jeog because there wasn't much in it.) I passed a big window on my left, which was a room for management. The lights were on in there, but nobody was in the room. The rest of the area had that gray, stock room atmosphere that is the same in any kind of retail operation.

Just beyond this was a standard door and a lighted break room with vending machines, refrigerator, cupboards, digital ovens, and four white, round tables with black, metal chairs around them.

Seated at one of the tables, all by herself, sipping artificial fruit juice from a tiny box with a straw, was Nola.

She had straight, black, shining hair, like a blanket, down to her shoulders, pale skin, a kind of exotic, Eurasian look, and she was wearing a green collared shirt like a golf shirt, and a gray skirt. Her knees were positioned together and her slender lower legs were angled sharply to one side, tapering to her tiny white socks and black pumps, ankles crossed below the table.

She did not look up when I came into the room.

"Hello," I said.

Without changing position, her eyes drifted up from the box of artificial fruit juice. She continued sucking the straw.

"You must be Nola," I said.

She stopped sucking.

"How do you know me?"

I jerked my head in a rearward direction. "Joeg told me."

Nola rolled her eyes. "Joeg."

She didn't elaborate. I felt awkward.

"Anyway, my name's Jeff. I just started in the liquor department."

She nodded slightly and returned her attention to her drink.

She was quite beautiful. I guessed that was why she was so chilly, and I thought, who needs her. I turned to the vending machines. I had a water card which was good for the machines.

Suddenly, Joeg came bouncing through the door in a burst of nervous energy, like a pheasant scared out of a bush.

"Ah! I left my bag in here," he said, his eyes slowly glancing all around the room, as if he might have left his bag on top of the refrigerator or nailed to the ceiling. "Now where did I leave it?"

"I see you're still here, Nola," he went on, smiling at her.

She raised her eyebrows and nodded.

"I guess you met Jeff. Did you meet Jeff?"

"I just introduced myself," I told him.

She nodded.

I noticed that Nola's eyes shifted toward one of the adjacent tables. A black backpack lay on one of the chairs beside it.

"I guess you're getting to know each other," he went on, unnecessarily. "Did you tell Jeff about your disappointment, Nola?"

Nola straightened. Her legs shifted from their slanted angle and made a straight up-and-down line to the floor. She drew in a breath and replied in a solemn tone that scarcely disguised her impatience.

"What disappointment is that?"

"You broke up with your boyfriend, I mean."

I knew better than to watch them at this point. I turned to the vending machine and became intimately concerned with it as Joeg floundered in this faux pas. Triangulation is a good method for getting the attention of an interesting woman, and the best way to do it is to shut up and mind your own business while the other guy makes a donkey of himself.

"I don't discuss my personal matters with people I've only just met," Nola was saying, a sharp edge in her voice.

"No, of course not. No, no," Joeg stammered.

"Is that your bag right there?" Nola said, nodding toward the black backpack on the nearby chair.

"Yeah, that's it."

Joeg retrieved the bag and acknowledged that, well, he had to get back to work now, and he would "leave us two together" — another terrible choice of words, but one that would not necessarily help me, I knew. So as soon as he staggered away, I grabbed up my stash from the machine and said to Nola, "It was nice to meet you. Excuse me," and I headed for the door.

"You don't have to leave," she said.

I looked back over my shoulder. "Excuse me?"

"You don't have to leave because Joeg embarrassed me."

"Well, I—"

"If you have somewhere else to go, that's fine, but don't leave because of me. The break room is for all employees. Did you have somewhere else to go for your break?"

"Well, actually, no. I—"

"Then come and sit down. Don't be ridiculous, like Joeg."

I sat at the table adjacent to hers. The same table would have been too much.

We talked for a few minutes—small talk, introductory matters. Then she said her break was over and went back out onto the floor.

I already felt myself going under her spell.

43

If You Think, You Go Crazy

Right about this time, and after a long spell without communication, I got a paper letter from Tom. It was postmarked Columbus, which is not where his prison was. I opened it up and read:

Jeff,

I'm in the hospital here in Columbus. I got beat up pretty bad and they sent me here. If I die it'll be a good thing. Now that Irene and Tommy are gone I don't want to live anymore.

Prison is the worst place in the world. It is hell on earth. The only way to adapt to it is to deaden your mind and just go from minute to minute and don't think about anything. If you think you go crazy.

It was another inmate that beat me up, I won't go into the details. Just an asshole. They came down on his ass too.

Anyway they sent me here and it's like heaven. Doctors and nurses and people who are real human beings. Some of the inmates are real human beings too but a lot of them aren't. So I'm here now and I hope I stay here and if I'm lucky I'll die.

Reason I'm writing is, I hope you're fine. You're a good friend, Jeff. I wondered if you heard anymore about the DNA thing. Do we just give up on that?

Well I hope you'll write me soon. I don't know how long I'll be here. If you could visit it'd be great but I know you don't have a vehicle.

Take care man. If anyone steals your water card, report it right away.

Tom

Reading this letter, I felt I'd been remiss, because I hadn't told him I'd hired Cyril Shaker. I didn't want to get his hopes up, in case Shaker failed to come up with anything. But now there seemed no point in concealing what I'd done; Tom was already despondent and couldn't go much lower. So I wrote him a long letter, and told him all about it.

The only thing was, I hadn't heard anything from Shaker, so there were no results to report.

It was after I mailed the letter to Tom that I got a skype from Shaker.

He was in Washington, D.C.

44

Shaker to the Tenth Power

I was all excited because I had a date to watch vids with Nola at her place, and I was getting ready for that when my media announced that "YOU HAVE A CALL FROM CYRIL SHAKER. YOU HAVE A CALL FROM CYRIL SHAKER. YOU—"

I was in the bedroom, finishing dressing. I flicked on the remote media control from the bed cell, called out, "Yo! Hold on a sec," and hurried down the hall to my major media room where Shaker was skyped in in multiple images around the room: wall vid, PC, tablets, laptops, miniatures. It was Shaker to the Tenth Power.

"Mr. Shaker!" I called out, confirming the connection.

He looked not so good, like he'd been drinking, but his speech wasn't slurred. Instead, he was red-eyed, bags under his eyes, weary-looking, rumpled clothing, and my second impression was that maybe he hadn't had enough sleep.

"Are you alone?" he asked me.

"Yeah, sure. I'm in my house. What's happening?"

"I'm in D.C.," he said. "I've got some information for you."

"Shoot."

"Jeff, the woman who was murdered. Her name wasn't Hypnotica Christiansen. It was Teresa Pagano. She was a secretary for your uncle, Preston Keys. She worked for him on a part-time, as-needed basis when he was doing his contract work for the government."

"What!"

"Yes. She created a fake ID for herself. She went to great lengths to do it before she left D.C. for Ohio. She was apparently very loyal to your uncle, Jeff."

"She worked for him?"

"Yes. This is good information, Jeff. Listen. This is very important. I want you to take very seriously what I'm going to say to you."

"I'm listening."

I was listening, but it wasn't easy, because my heart had started pounding very hard in my chest, and blood was coursing loudly through my ears, and I had trouble controlling my breathing.

"Jeff, you may not be safe. You must be very careful. Are you listening now?"

"Yes, sir."

"Good. I'm not positive about this last bit of information, but I think it's right, and it's what makes sense. I believe Teresa Pagano had copies of the vids that you showed me, and others besides. I believe she had them on her person when she went to Ohio, and that she was killed by someone who wanted to retrieve them from her. Nothing of that sort was found on her body, or in her possession, of course."

"What are you saying?"

"Jeff, I'm saying that there are"—he lowered his voice, actually whispered the next two words—"government agents who are looking to retrieve those vids, and will stop at nothing to get them. I'm assuming that Pagano was in Wooster to take the vids to you."

"To me?"

"There's no other reason for her to have been where she was when she was killed. And there's no other reason why the kid who robbed Tom would know your name."

"You mean that I'm being watched?"

"Watched, followed, monitored, you name it. They got the vids from Pagano when they killed her. They may believe they've got everything they were after, but on the other hand, we don't know that for sure. So I'm telling you, until I can ascertain that for certain, be careful about getting into strange vehicles.

Don't respond to invitations from strangers. Don't go out for evening walks by yourself in lonely places. Most of all, be careful who you hang out with."

I said I would be careful—but immediately, I naturally thought of Nola, whom I didn't know all that well—because she was the only person I was planning to 'hang out' with at all at this point in my life. Yet there was no way I was going to pass up our date tonight. I thought it best not to tell Shaker about her.

"One other thing, Jeff—and I leave this up to you."

"Yes?"

"Have you thought of arming yourself?"

"What, you mean, buy a gun or something?"

"Yeah, something small and concealable. Apply for a permit. Like I say, I leave it up to you."

"I'll think about it," I said.

"OK, Jeff. That's all for now. But I'll probably be calling you again soon with more information. Please be very careful. Eyes in the back of your head. Understood?"

"Understood."

But I did not fully understandand I was soon to find out just how costly that incomplete understanding was.

45

I Hope You Like Rice Casserole

Soon after I'd received the fifty thousand dollar check from my mother, written on my uncle's estate, I'd updated my bicycle by installing an electric motor on it, effectively converting it into a moped. This would make it easier for me to get across town when I needed to, and wouldn't be nearly as expensive as buying an Electric Vehicle or a Wind Vehicle, which would wipe out my little nest egg.

That little update came at just the right time too, because Nola lived uptown in a trendy little condo, not all that close to my downtown location. I charged up my bike battery and pedaled off, letting the motor kick in on the long stretch of Cleveland Road, which was one of the main north-south drags, and that saved me plenty of time and energy.

This was my first date with Nola, and it was a natural thing that we should be seeing each other, because we got along well at work, and we had in common the fact that we had each recently broken up with our more-or-less-significant others—and, as it turned out, for the same reason. She said her boyfriend, George, had wanted to "get serious" and "take their relationship to the next level," but Nola was not entirely convinced that she had wanted to go to the next level with George. Like me, she would have been content to continue as she had been with her S.O., but that hadn't been enough for him. I guessed Kareena was right: it either gets better or it gets worse.

Nola's condo was in a cul-de-sac, with pleasant large pine trees overshadowing the nest of identical residences sitting side-by-side, with that kind of creepy uniformity such buildings have—in this case, pink brick and white sidewall. I pulled into the white driveway on my bike and walked it up onto the porch, opened the screen, and knocked on the oaken door, which had three sunburst lights in it.

"Come in," came Nola's call.

I stepped inside, and there she was on her sofa, underneath a lighted lamp, reading an antique paperback, not looking up. She was wearing brown corduroy pants, a peach-colored light sweat-shirt and matching socks, no shoes. Her toes were toward me as she lay reclining and they wiggled absently.

"Hey, listen to this sentence and tell me what you think," she said.

I waited. She read aloud:

"It was their first date. They hit it off immediately and soon they were making out on the divan. She sucked his tongue with passionate promise."

Nola looked up, awaiting my judgment. I could see from the book's cover that it was a romance novel. It had a guy with his shirt torn off and hanging from his belt, and the guy was passion-ately pulling a young female against his loins, his head poised above hers, like an eagle over a mouse, as he prepared to close the oral gap. The woman, head bedecked in long cinnamon ringlets, was in some kind of period-piece get-up with lots of petticoats, but her décolletage was highly revealing and the petticoats were in the initial stages of hike-up.

"That's more than one sentence," I said.

"What? Oh, I know it's more than one sentence. But what do you think? Is it over the top? Tasteless? Or does it work? I mean, given it's not War and Peace?"

"I think it needs some details. 'It was their first date. They hit it off immediately.' That's not vivid."

She set the book on her end table and sat up, casting a caustic glance at me.

"I hope you like rice casserole."

I sensed that the passage she'd read was over the top because it referred specifically to a first date. That's why I deliberately played it down by not referring to the passionate promise. Something was suspicious, even if I hadn't been made paranoid by the skype from Cyril Shaker.

We ate the casserole and watched the vids. Nola turned out not to be at all like the cool character I'd first met in the break room. I learned that she was from Baltimore, that she'd come to our fair town to take a few classes at the college after having broken up with her boyfriend, but that hadn't lasted, and she ended up working at the market as a stop-gap measure. I didn't learn a tremendous amount more. Pretty soon we were making out and I became even more suspicious when she said it was OK if we went to bed.

In my experience, that wasn't something that usually happened on a first date. Still, I wasn't going to turn it down.

46

The City Was Burning

The next morning was a Saturday. I turned on my AllVids to catch the replay of Wesley Wright's midnight broadcast, when I came across the news programs. They were saying the fires in California had spread into San Francisco after a huge earthquake in the wee hours.

Fires in California were nothing new. They were the norm, and since there was no water there, you were always hearing how they'd burned breaks in the line to try to stop it here or there. The mountains were burnt to a crisp. But the last I heard, the Northwest Coast fires hadn't gone below the Russian River, which on the maps is north and east of San Francisco.

Now the city was burning, they said.

In retrospect, it was a good thing the population there had dwindled over the years, as it had in L.A. and the other big cities out there, but there were still plenty of people, and now they were in for it.

It was the Big One.

I called my mother to see if she'd heard about it, but I got no answer.

Saturday was a work day for me and I went in at nine o'clock, feeling suffused with well-being from the joyful workout of the night with Nola, who was not in the store on Saturdays.

Joeg Hibben was.

I passed him on my way to the time clock. I said hello, but instead of replying, he cast a look at me from that asymmetrical face, like Bela Lugosi confronted with a crucifix in those antique films from the twentieth century. I thought I almost heard a hiss as he cringed and slunk away in the opposite direction.

It was as if he knew what I'd been doing for the past twelve hours, but I was sure that all he knew was that Nola and I were friendly at work, and conversed easily—much more easily than he did with her. That was enough to make me anathema to him.

Well, what could I do? The law of life is that where there are women, men will be in conflict over them. Therefore, a man must be prepared to be hated by other men, if he has any luck with women at all.

I felt so lucky, I didn't want to jinx it by asking myself if I were in love, or if Nola was maybe a spy sent to keep tabs on me, along the lines of Shaker's warning.

Speaking of which, I also had no plans to buy a conceal-and-carry weapon. It wasn't that I didn't believe Shaker about the danger I might be in. I did, and that was another reason I had wanted to talk to my mother. I was concerned that the danger could spill over in her direction. But I couldn't see myself pulling a pistol out of a shoulder holster from under my coat and blasting away at somebody. I didn't want to kill anybody. That kind of stuff was for the holos and the movies—blast away and never think twice. No repercussions. That's fantasy. I was scheduled to get off at six that evening. But around five o'clock, I was out on the floor stocking bourbon pints when Mr. Barrington came walking up, surrounded by three cops in full dress uniform. They headed right for me.

"Jeff, these officers need you to go with them," said Mr. Barrington. "Take the rest of the evening off."

"What is it?" I said. "Is something wrong? I'm not in trouble, am I?"

"No, sir. We just need you to come with us."

I clocked out and went with them, but by the time we hit the parking lot, I figured out it had something to do with my mother, and I said so.

"Yes, sir," they said. "There's been a crime."

I was in their cruiser by the time they led up to telling me what had happened.

"Mr. Claymarker, your mother appears to have been killed by an intruder into her home. We need you to identify the body."

47

Getting Really Gestapo

They didn't need me to identify the body.

What they wanted was a confirmation.

I did that in the morgue. It was a ghastly place, with sheets laid out on tables and bodies under the sheets. There were rows of metal tables like bunks, and rows of lockers, and I felt like I was in some kind of nightmarish butcher shop reeking of formaldehyde. They brought me to a body with a tag on the toes sticking out from the end of the sheet, and they lifted the top of the sheet and pulled it back about a foot. The sight of my mother's face horrified me. It was discolored, and her tongue was sticking out. I got immediately sick, but I guess they knew I would, because there were big white-coated Techs on me right away, catching the contents of my stomach in buckets.

Then I passed out, and woke up in the E.R. at Wooster Community Hospital. They were hydrating me with an intra-venous tube, the old needle taped against the arm and stuck in a vein. A nurse was standing over me.

"How are we feeling?" she asked.

I refrained from commenting on her first person plural.

I told her I was OK, and asked where my clothes and wallet were, because I wasn't wearing them. The nurse smiled and tapped a locker next to the room's door.

"Locked in here. I've got the key. It'll be kept at the desk for you when you're ready to claim it."

After she left me, I did some crying. I kept remembering that my mother hadn't answered the phone when I'd called her in the morning, and if I had checked on her—then I looked around for a clock. Was it still Saturday? What time was it?

When the nurse came back, she asked if I wanted a late

dinner, so I asked her what time it was.

"Eight thirty," she said.

"Not hungry."

She left, and I eventually fell asleep, but only for about three hours. When I woke up, it was after midnight. I had by now spotted the digital clock on a nearby shelf.

I couldn't sleep. I kept hearing the hospital staff talking outside my room, making no attempt to keep their voices down, and it was like listening to an endless lunch break conversation until I heard somebody say something I hadn't heard before, about the news of the day.

"The Texas legislature. That's who."

"Based on what authority?"

"Who knows? I guess they just decided they could declare independence."

"Is that constitutional?"

"That's not the point. They're declaring themselves independent of the constitution."

"But that's been tried before. It doesn't work."

"Times have changed. The states have become a drag on the federal government, especially down in the Gulf Coast, with all the flooding. Just think of all the federal emergency money they could save by letting Texas go."

"They'd still send money. People have relatives in Texas. We wouldn't just let them go under water."

"Why not, if that's what they want?"

This conversation was composed of three voices, two male and one female. You could tell they were intelligent voices, and every now and then I could see one of the speakers through the cracked-open door to my room. It was a young man, with a green face mask tied around his neck with a little bow.

"Goya will send in federal troops. Texas won't have a chance."

"Are you kidding? This is Goya's big opportunity. Just think how much better off he'd be in the '52 election without Texas in the mix. They hate him down there."

"He can't just let them go, though. It'll start a ripple effect."

"Look, the feds can treat all the refugees as foreigners if they let Texas secede—the ones from Texas, I mean. They can deport them. That'll take a lot of pressure off the state governments and the federal welfare subsidies."

"Did you see they're introducing a state law—"

"In Texas?"

"No, in Ohio. A state law that's going to require every home with an extra bedroom to house a refugee."

"They can't do that. Is that constitutional?"

"Oh, they can do it. They are doing it."

"What do you mean an 'extra bedroom'?"

"Any bedroom without someone sleeping in it, is what I understand."

"What if I use it for something else? What if I keep my sound equipment in there?"

"Too bad. If it can be used as a bedroom, you'll have to have a bed in it, and a refugee in the bed."

"This is getting really gestapo."

"The refugees are getting out of hand. They've got to do something."

"That's what they always say before they do something stupid."

"Well, I think they ought to let Texas go. For that matter, we should dissolve the union. It's not working anymore."

"Nothing's working anymore. It's not just the union."

"You got that right."

The conversation tiptoed on. The hospital staff seemed to have no work to do. I listened to them, against my will, all night long, because I didn't sleep. In the morning, a detective from the police department wanted to talk to me, and the nurse asked if I felt up to it.

48

I'm Sorry About What Happened to Your Friend

I told the nurse I was feeling perfectly OK now—just depressed over the loss of my mother.

"Do you want to go home?" she asked me, and I said I did, so she said she'd be back in in a few minutes to do one last check of my vital signs, and then she'd release me. In the meantime, here was the detective.

In came the same guy who had interviewed me about Tom's water card. I didn't remember his name, but the face registered.

"We meet again, Jeff," he said, smiling as if we were old pals.

"Yes, sir," I said.

"Hayes. Detective Hayes," he reminded me.

"Yes, sir."

"Jeff, I'm sorry about what happened to your friend. I know you didn't think he was guilty. And now I'm even sorrier about your mom. She definitely didn't deserve what she got."

"No, sir."

I looked away, signaling to him that both subjects were distasteful to me.

"Reason I'm here, Jeff, is—when you feel up to it—we would like you to go through your mom's home and identify missing items for us. We don't know what all was in there, but it appears some items were taken. There aren't any computers in there, for example. And of course the A/V stuff is missing. You know, burglaries are usually drug addicts, and the first thing they take is the A/V, and then the computers. You know, TV screens, stuff like that. But some of the drawers were not even ransacked. Her wallet and purse weren't taken, and a considerable amount of cash was found in a drawer. We're not sure why they wouldn't have taken it, unless maybe someone surprised them when they weren't done, something like that. But if you could come to the house and go through it. I assume you're familiar with what she had in there?"

"Pretty much."

I didn't know that I could identify every last missing item. Think about it: do you know where your mother keeps all her things? But I knew the basic layout of the house, and where all the big items were.

"So, you think you could help us out, Jeff?"

I nodded.

"That's great," he said. "Listen. When do you think you'd be ready to come down to the house?"

"I can come today, if they let me out of here."

He was clearly pleased with this reply.

"That's great!" He fished a card out of his shirt pocket and handed it to me. "Call me on my cell before you come down. The place is all taped and cordoned off. You've got to be with me to get in there. I'll come and give you a ride, if you need it."

"All right."

As he was backing away, he paused, and adopted a thoughtful tone.

"Jeff? You know, when we prosecuted your friend. There were some guys on the force that said, 'That friend of his was in on it.' Meaning you, Jeff. But I told them, 'No, he didn't know a thing. He went home. We've got nothing on him.' I just wanted you to know that."

I couldn't bring myself to thank him, so I simply nodded — with eyes closed, as if I were a little tired.

"Well, I'll let you get a bit more rest here, Jeff. You call me soon's you get out now, you promise?"

"Yes, sir."

"All right."

His eyeglasses flashed as he backed out of the room, passing the nurse on her way in. I was out in half an hour, and I had no ride back downtown, so I called the detective right away. He took me out to breakfast—pancakes, maple syrup, sausages and coffee, no wafers or gruel. Then we drove out to my mother's house for the inventory.

49

Crime Scene

My mother's blue, vinyl-sided three-bedroom bungalow sat on a street near the railroad tracks in Rittman, behind a green lawn and a white ash tree with leaves almost purple in their full autumn bloom. Yellow police tape cordoned off the yard and front walkway, and as I was carefully led by Detective Hayes and two uniformed officers to duck underneath the tape, I saw the words "CRIME SCENE" in big black caps standing out against the yellow tape, and once again I felt like throwing up—but managed to hold it in.

"Don't touch anything," Detective Hayes directed, "but, as you walk through, call out the name of any item you think should be there that isn't, OK?"

He had his marker and electronic notepad at the ready.

"OK," I said.

First up was the living room. That was easy, because you could see the holes in the entertainment center, and the loose cords trailing into the walls.

"TV screen."

"Check."

"Remotes, router."

"Right."

A small, portable digital music player still lay on top of a rolling stand with drawers.

"I don't know why they didn't take that," I said, pointing to the player.

"We don't, either," Hayes responded.

The drawers to the rolling prop were standing open, as were a number of drawers beneath a built-in bookcase in the corner of the room. Half the books had been swept onto the floor, and a few lay on chairs at that end of the room, pages askew, but it looked as if all the books were still there. There was no damage to lamps, sofa, the three occasional chairs, which were all cushioned swivel chairs. Some of the throws were out of place or on the floor.

I peered into the open doors of the player stand and saw that it was completely cleaned out.

"This wasn't empty before," I said.

"What was in it?"

"I'm not sure. I think it was mostly music. A lot of old CD's, but then that was replaced with tiny digital storage units, like before I was born. All that's gone."

"OK."

Hayes was busy keying on his pad.

"What about these drawers under the bookcase?" he asked.

I stepped over the fallen books and took a look through these drawers, which were all pull-outs.

"These would have been full of stuff, but I can't say exactly what. Photographs, albums, old-tech kind of stuff. Papers. Just a lot of junk my mom collected over the years, or that she inherited from her parents and grandparents. It's all gone, whatever was in there."

"OK, so like, memorabilia, photos."

Adjacent to this room, and contiguous with it, without doors, was the kitchen. There, the drawers had been similarly ransacked, but not entirely cleaned out. Phone books and notepapers were all over the floor and on the kitchen table, but telephones were missing.

"And also address books," I added. "I'm sure she had some address books in these drawers."

"Address books," Hayes repeated, writing it down under "phones."

"That ain't like a drug addict," one of the uniformed cops grunted, and the other one shook his head.

Nothing else in the kitchen was touched.

We then proceeded down the hallway into the three bedrooms. It was the same story. Telephone and audio-visuals missing and drawers ransacked. Clothing was untouched, and Detective Hayes said he'd already removed jewels, a purse, wallet, and cash that had been left behind, and had sent them to the evidence safe at headquarters.

And that, I thought, ain't like drug addicts, either.

My mother had converted one of the bedrooms into a den. Here, she kept her computer equipment, all old stuff that hardly anyone uses anymore, but it was entirely gone and easy to recall and identify. All the peripherals were missing as well.

"Yep, yep," clucked Detective Hayes, writing this down as if it were entirely expected. "All computer materials and peripherals."

The third bedroom was a guest room. I knew that room pretty well, because it was where I would stay from time to time. This room was almost entirely taken up with a fold-out bed, but there was a nightstand, two drawer chests, and one other significant item. The rooms were all carpeted, and as I entered the guest room, I immediately sensed something different.

It was a space that hadn't been there before. It was next to a ransacked chest of drawers, from which nothing had been taken, because that chest had contained only sheets and towels. But the space next to the chest was now a deep, rectangular indentation in the beige carpet.

"File cabinet," I said. "There was a file cabinet right here."

I formed the shape of the cabinet with my hands to show the detective.

"File cabinet," he repeated. "How many drawers?"

"I think, four. A vertical cabinet."

"Color?"

"Black, or dark brown."

"And what was in it, you know?"

I shook my head. "Just all her important papers, I think. Anything you'd put into a file cabinet. Bills, insurance policies, certificates of title for motor vehicles, tax records. I mean, I'm guessing now."

"Important papers."

"This don't make any sense," said the uniformed cop who had spoken before. "Computers make sense. But tax records don't make sense. What do you think, Captain Hayes?"

Hayes frowned, brow wrinkled over the top of the eyeglasses.

"Bit of a puzzle. You've got the A/V and computer thefts, which is like a typical burglary. We've already got men out at the pawn shops looking for that stuff. But why the thieves left the cash and jewels and took the address books, you got me on that one. It's almost like they wanted information as much as value. Got any ideas, Jeff?" he said, turning to me.

I shook my head. I could feel my own pulse at my shirt collar.

That was a dishonest response.

My head was full of ideas.

50

The Only Two People on the Planet

When I finally got back to my house, I went to my bedroom, but instead of lying down, I sat on the bed and phoned Nola.

"Jeff! Are you coming over tonight?"

I told her, speaking quickly and pointedly, what had happened to my mother.

"Oh my god, Jeff! Are you serious?"

Her exclamations kept coming, and I listened to them intently. I was trying to gauge the precise degree of actual versus feigned surprise.

I can't prove it, but this is something that I've always felt I'm good at: gauging whether someone is sincere or not. Like when you watch a politician and he says some words which are intended to inspire support. If the words are insincere—if they are delivered solely for political purposes—you can tell. The words hang in the air, and you see the face of the politician below them, and the face of the politician shrivels up into ashes beneath the words, and you're left with empty, fake words.

I've seen this many times.

I'm not saying I can't be fooled. A good actor can fool you. But a bad actor can't; and most people are not good actors.

So, as I listened to Nola tell me how shocked she was to hear my mother had been murdered, I gauged it, and the result was mixed. There was some genuine shock. But there was also a strange lack of convincing surprise. It was like the surprise was acted.

So, shock without surprise? How does that work?

Think about it. It is always shocking when someone is murdered. Even if you know it's coming.

"Jeff," said Nola, "you can come over here, if you want. You need comforting. I can help you."

"Yeah, I want to come over, I really do," I said, "but not just yet. Give me a few days."

"Of course, Jeff. Of course. Take as long as you need. I understand perfectly."

Next I tried calling Cyril Shaker. He was the person I most wanted to talk to right now. Because I told you before: I'm not dumb. And if spooks had set up Tom, and if my Uncle Preston had had the information that explained how they had done it; and if shortly after Uncle Preston's death, my mother was his next of kin; and if she were then murdered; and if what had been stolen from her house was every form of information recording or purveying device, as well as personal information of no monetary value, while no other valuables were taken—if all that was true, then it was clear that what the thieves wanted was information that had been in the possession of my Uncle Preston.

And right now, so far as I knew, Cyril Shaker and myself were the only two people on the planet who had this information.

51

Piece of Cake

The story of my mother's murder was in the online and paper news reports the next morning. I got calls from people I knew, like Grady Lurie, who said he was really sorry to hear about it; and from Mr. Barrington at work, who told me to take two weeks off. I was only entitled to one week with pay, but he said he'd make it two.

Then I got a skype from a lady who claimed to be a friend of my mother's. She looked to be about my mother's age, or a little younger. She had the Brazilian coloring that most Americans have, and curly, iron gray hair. She wanted to know when the funeral was.

"I haven't made those arrangements yet," I said.

"Will it be in the news vids?"

"I hadn't really decided, to tell you the truth."

"You've got to put it in the vids or the papers. That's the way things are done."

After she hung up, I felt like throwing my handset at someone. Here were strangers telling me how things were done, and they had no idea that advertising my mother's funeral (and therefore where and how to find me) might lead to my own funeral.

Still, I felt I had to tell friends and relatives. There were only a very few to call because the only Claymarker relative had been a sister of my mom's who had been dead for six years, and there were some cousins that we didn't even know down in southern Texas. But she had a few friends whose numbers I found among some records I had and I called them.

One of these was a platonic boyfriend named Fred Grier, who broke down and wept right on the phone when I told him. But then he gathered himself and told me I needed to contact the attorney she'd been working with on Uncle Preston's estate— which was true, because now it would all come to me.

I made arrangements for cremation, and the funeral director I dealt with tried to strong-arm me into a big funeral with notices in the paper and vids, but I resisted. He was Not Happy. This was Not The Way Things Are Done, he wanted me to know.

On Monday night, I finally heard from Cyril Shaker again. I was in bed when he called at about eleven p.m. I reached for my wrist vid, which I keep on the nightstand by my bed, and received his tiny skype while blinking myself awake in the dim low beam of a gradation touch lamp.

On the tiny dark square, Shaker looked haggard and exhausted. He seemed to have aged five years. All I could see was his bespectacled, professorial face and one of his hands as he gestured with it. Purple-blue shapes of furniture in the dark background, a yellow glare from a lamp. Despite his worn appearance, he seemed enthusiastic, eager to give me his latest news—almost giddy.

"I'm wrapping it up here in D.C.," he said. "I'll be back in town by tomorrow night about this time. Listen, Jeff. What the spooks were after is one particular file of your Uncle Preston's. I don't have the label. I don't know how he identified it, but it's bound to be among those hundreds of digital files you copied for me, which means I've got it, and you've got it, most likely. It's something to do with a closed Senate subcommittee hearing that took place in 2025. My information is that Teresa Pagano had a copy of it on her when she was killed. That means they got it back, but they may or may not know other copies exist. For that matter, we can't really be sure there are copies until we find them. I'm going to do that search when I get back home."

"How are you getting this information, Mr. Shaker?"

"Plenty of time to talk about that later, Jeff. Short version: People have friends. And Pagano, she kept a red tablet. She gave it to her friend before she left for Ohio. I've got it. Haven't reviewed it yet, but no one else has got what I've got. There's only one tablet."

"How will all this help my friend, Tom?" I asked. That was, after all, the point.

Shaker laughed and shook his head, as if I had asked a question of little significance.

"Don't worry about Tom, Jeff. Tom is in good shape—great shape. I'm not worried about Tom."

"But what do we do when you get back? How do we get him out of prison?"

Now Shaker began to grow impatient. "It's easy. We hire a lawyer. He files a post-conviction petition. We might not get the conviction thrown entirely out, but there's no question we get him a new trial. Then the lawyer puts me on the stand. I'm your star witness. Piece of cake, Jeff. But what we've got here is something bigger, much bigger. We're sitting on a nuke, Jeff, and we've got to be real careful. Oh, damn. Damn, damn, damn! Sorry, I'm out of minutes, Jeff. I'll call you back tomorrow."

The tiny screen went black.

At 3:30 a.m., my media went off with another call from Shaker. There was no vid this time, only sound.

"Hello," I answered. "Mr. Shaker?"

There was no answer, just a creepy, crawly, scratchy sound that made me think of a cockroach crawling over a pile of debris. I don't know why it made me think that, because that wouldn't make any sound at all.

"Mr. Shaker, are you there?"

I heard a whisper that began at the tail end of my query, so I didn't make out what was said. Then it repeated itself, more forcefully.

"Five-thirty," it said. "Five-thirty."

Twice, just like that.

Then I heard interference, as of hardware clunking around, and I heard a voice in the background. I couldn't make out the words, but it sounded like the exasperated initial consonants of swearing or cussing: like J—or Shh! Then I heard a small, muffled pop, and the line went dead.

52

Five-Thirty

I had so many questions after the 3:30 call that I couldn't go back to sleep. When I tried calling Shaker, I got a disconnection notice. After a couple of attempts, I realized it might not be a good idea to be trying to call him, and that I was maybe lucky he was disconnected.

Here is why I thought that:

First. The last thing he had told me in the 11:00 p.m. call was that he was out of minutes. So how does he call me back at 3:30?

The only answer to that is that he realized he was running out of minutes during the 11:00 p.m. call and he cut it short with maybe a minute left. Then, at 3:30, something happened that made it vitally important to reach me with the last minute he had left on his media.

Second. When he got me with one minute left, he said the most important thing he wanted me to know: "Five-thirty." He said it twice, it was so important, but he didn't say anything else. I thought: He was partially incapacitated.

Third. Voices in the background cursed, then I heard a muffled pop, and then the line went dead.

The only thing this meant to me—and I desperately didn't want to admit it to myself, but could think of no other explanation—was that spooks had killed him after his last word to me.

Five-thirty.

I was terrified. What did "five-thirty" mean? Did it mean spooks would be coming for me in two hours? Did they know where I was? Were they going to kill me, too?

I had to get out.

I sprang out of bed—not at all sleepy now—and threw on a pair of jeans, a shirt, sweater, tennis shoes, and a jacket. I gathered my wallet from a desk drawer, my house keys, locked up the house, and went out for a walk.

I planned to be gone until well after five-thirty a.m.

Oh, I almost forgot.

Fourth. If spooks killed Shaker, I didn't necessarily want my Caller ID showing up on his media right at the moment.

53

Spooks

Spooks.

Government people. Clandestine operators. Secret black ops. Dirty tricks.

Tom and I used to see all the vids and read books about them. They'd always been there. But Tom said they rose to their present level of power after the Second World War in the twentieth century, and they worked for the big corporations and the super rich. Oil and gas men. Frackers. Coal. Mining operations all over the world. Bananas in Central America, coffee, sugar. Chocolate. Rubber. Hemp. Resources. They ran the show, and presidents and congress people were like dancing marionettes in their hands. If the marionettes became too troublesome, they were rubbed out like so many tally marks on a slip of scratch paper and the tracks were covered over. Spooks controlled the media, controlled what people think, controlled the narrative.

They controlled history.

We all know this. We, the little people who are not so dumb.

There's a lot of us, really. More than you might think.

And I had the evidence of spooks in action right in front of me, with what happened to Tom, and what my uncle's vids had on them.

So I was properly scared when I went out for my three hour walk to make sure I wasn't home at five-thirty.

A funny thing happens, though, when it gets to be six-thirty. You get physically tired of walking up and down the streets, looking for something to keep you occupied. And you get tired of being afraid. You want to go back to your home. You think, all right, if they want me so bad, let them get me.

I reached this state of mind as the light came into the cloud-crammed sky—such light as we ever have—and I was sitting on the grass, out by the pond at Schellin Park, which is a postage stamp sized piece of lawn near the freeway entrance ramps, leading into the downtown area. A guy about my age came walking up, oblivious to the world, on the other side of the pond. He sat down and tied up his left arm with some kind of rubber cord, sleeve rolled up, and pulled a syringe out of his pant leg and shot up right there in plain view.

I'm going home, I told myself.

I hiked under the overpass in order to avoid meeting the addict and crossed the weedy, cracked pavement in front of the old shuttered homes and high-rise apartments, the post office, and then the town square, heading for my street.

The house looked quiet, parked behind matching shrubs. A modest little house with brick and blue siding, nothing out of place or suspicious. I had a little bit of money now, but I had been poor a long time. And I owned no vehicle, other than my moped/bike. And right now, owning no vehicle seemed like an advantage, because it meant that no vehicle of mine could be rigged to blow up when I got behind the wheel.

I couldn't see any sign of spooks about the house anyway.

The fact is, I had only been guessing at what Cyril Shaker might have meant when he croaked out the word, "Five-thirty" in what sounded like his last breath. The fact is, I had no idea what, if anything, had really happened with Cyril Shaker. I was eager to find out.

In the meantime, I came to an agreement with the funeral director to do a small, quickly organized memorial service for my mother, without her body being present. We got notice out on the vids same day and held the service two days later. I was surprised at how many people turned up, mostly elderly friends of my mother, but some younger ones too. I was so busy with that that I had little time to surf cyberspace or catch any vids. And as I was leaving the funeral home, I saw an old-fashioned paper news rag half-crumpled and lying on a chair, and something on the front page caught my attention. I stopped and snapped up the paper, heart pounding.

It was a photograph of Cyril Shaker's face as a considerably younger man, below a headline that read, Local Ex-Cop Kills Self in D.C.

54

Not Worth a Get Out of Jail Free Card

The story was a couple of days behind the facts. Shaker had evidently worked as a policeman here in town for many years, before becoming a private investigator, and this was the angle that interested the local reporter.

His body had been found in a hotel room in D.C. with two bullets in him. The story did not describe the location in his body of the bullets, but said the D.C. coroner had ruled the death a suicide. An automatic pistol was found near the body with five shells still in the clip.

I knew damn well Shaker was not a suicide. The pistol might have been his, or it might trace back to someone else, but it wouldn't matter, I knew. The information I was interested in was not in the news story, and would never be in any news story, namely: where were the media recordings I had given to Shaker? Most of them had been on miniature media, so it was possible that he had them with him, or he might have left them back here in Wooster.

The next day I got a call from Shaker's wife, who said she understood I had some funds on deposit in her husband's bank account and she would see that I received a proper refund as soon as matters could be sorted through.

"Do you know about the documents I gave him, where they are?" I asked her.

"No, I don't know anything about that. You can call his secretary about that."

I called the secretary.

"I honestly don't know without going through the files," was her answer.

It seemed that Unturned Stone had only two employees:

Shaker and the secretary. There was no backup private eye. The secretary said I'd have to hire a new detective and arrange with the new hire to examine my files. She could only identify my file and return it to me. I said that was all I wanted. That and the refund.

The spooks had won, just like they would always win. It seemed to me that nobody would ever spring Tom from prison. I believed then that he would live in hell for the rest of his life, for absolutely no reason.

And now my own life was looking like it wasn't worth a Get Out of Jail Free card.

There was one thing left I wanted to know before I was cornered and classified a suicide by the spooks.

And here it was a weekend coming up, too.

I called Nola, and she said I could come over and see her.

55

I Need a Contact

I had to walk to Nola's, because the streets were flooded and my bike couldn't have negotiated it. The water hadn't reached my house yet, but all the outlying areas of the town were under water: the fairgrounds, the fields and housing tracts bordering Highway 30, about half the downtown and some of the lower ravine streets. It was rain and thunderstorms all the time. Clouds as black as Dracula's cape, evil, red patches of light in between them, forked lightning whacking overhead.

It wasn't cold, for October.

I took an umbrella, but by the time I got to Nola's, it was rendered inoperable, and I was pretty well soaked.

"You'll catch your death," she said, as I dripped onto her inside welcome mat.

"I'll be OK."

She took the destroyed, sopping umbrella and put it in her bath tub, then got out her hair dryer and ran it over my clothing for several minutes, and that helped a bit.

"We've got to talk," I said, once the noise of the hair dryer had died.

"Have you eaten? I can make dinner."

She had a pair of tiny game hens already done in the oven. I was glad to get something other than wafers and gruel. We had peas on the side and bottled water. We ate at her kitchen table, with a folk art calendar on the wall and a ceramic chicken used as a sugar bowl.

"What did you want to talk about?" she asked, her eyes averted, not trusting the demand. Like all beautiful women, she assumed I wanted to get serious, and it was too soon for that, despite our recent intimacy.

"I have to tell you something about me that's not generally known," I said, choosing my words carefully.

"Oh?"

"I'm involved in something dangerous."

"Dangerous?"

She had her fork and knife in the breast of her game hen, slowly cutting the small amount of meat on those bones. She kept her eyes on the plate.

"I got myself into something that I shouldn't have, and I'm trying to get out of it. But there are powerful people that I've crossed, and I could end up being killed at any time."

"Surely not."

"I know it sounds outlandish, but three people have already been killed. They were people associated with me. And another one has been sent away."

"Sent away?"

I didn't want to get into the whole story about Tom just now.

"Did you see the story in the paper about the local ex-cop who committed suicide in Washington, D.C.?"

"No. What local ex-cop?"

"His name was Shaker. He was a private detective here in town. He was working for me when he was killed."

"You mean when he committed suicide?"

"He didn't commit suicide. That's just a news story. He was killed. People wanted him dead because he had information they needed to control."

"What kind of information?"

"I can't go into that. The point is, I've crossed a bunch of spooks, and I think I'm being followed, surveilled, watched. I may be a marked man."

After listening to all this with her eyes on the game hen, she finally raised her gaze to mine. The gaze was serious, skeptical.

"You realize how improbable this sounds, don't you?"

"Yes."

"So why are you telling me this? Do you expect me to believe it?"

"I don't care whether you believe it or not. But I'd like you to tell me more about yourself."

"About myself? What can you possibly mean, Jeff?"

"I asked you before where you were from. You said Baltimore."

"That's right."

"So what are you doing in a cow town in Ohio? How did you get here?"

"I think I told you that too. I enrolled in some classes at the college. I dropped out. I needed a job. I got hired at the market. Then I met you."

"You met me, and we fell into bed on the first date."

"Is that a complaint?"

"No, it's not. It was wonderful. I would like it to happen again."

"It could."

"But I'm trying to figure out how I got so lucky so fast. I mean, I was never told that I had such incredible charm—quite the opposite."

I was remembering how Irene had been fond of calling me a lizard.

"Maybe you underestimate yourself."

"Nola, do you remember when I told you my mother had been killed?"

"Of course. It was just last week."

I didn't say what I was thinking about that. I was thinking how she had not seemed truly surprised.

And there was only one reason why she would not be surprised.

"Listen," I said. "The truth is, right now I'm a person Interesting to Certain People. They have already killed or ruined the lives of four people associated with me. Call me paranoid if you want, but anyone who is very interested in me at this point in my life is likely to have the same reason as the next guy who is interested—or the next lady."

She was watching me now with those dark, semi-Asiatic eyes, and I thought I saw in them something like a newfound respect—and that clinched it for me. To put it bluntly, I believed she was one of the spooks. No, I knew she was. So the thing to do now was move in for the offer.

"I'm beaten," I said. "I want to deal. I don't want to 'commit suicide,' like Mr. Shaker, or be killed by a burglar like my mom. I never meant to get mixed up in this situation in the first place, and the only reason I hired a private eye was because I wanted to help my friend, Tom. But I realize now that nothing can help Tom."

I was speaking fast, intensely, not letting her interrupt me, not giving her a chance to break my momentum, because I wanted to get the offer completely on the table. If I was wrong about her, then she'd react with complete astonishment and consternation, and she'd get rid of me, as being a paranoid lunatic, and I'd never see her again. But if that didn't happen, we might have a deal.

"Listen," I said. "I know what I have that the people in D.C. are after. Shaker, the private eye, the dead ex-cop, told me what it was. I'm willing to turn it over to them. They can copy my hard drives if they want. They can have everything. All I want is to go on living, to have my life back."

I thought I might have her on the hook, but it was not yet a done deal. So I took the offer to the next level.

"Listen," I said. "The ex-cop's last phone call to me, he said, Five-thirty. He said it twice. Those were his last words to me. It was three-thirty in the morning when he said that to me. I was terrified. I thought he was telling me they were coming for me at five-thirty, in two hours. I got the hell out of my house and walked around in the rain for three hours, just trying to stay out of danger. Listen, Nola. I don't want to live like that. I need a deal. I'm willing to make it. I need a contact. I need to work with someone. I'm not going to blow their cover. I'm going to cooperate. I just want to be able to be in my own home at five-thirty in the morning without being afraid someone's going to come and force me to commit suicide. Can you understand that?"

Something in her eyes shifted, and I thought I saw a flicker of what I would never have expected to see — compassion.

"Jeff," she whispered. "About five-thirty?"

56

Banana, Tomato, Novice

"What about 'five-thirty'?" I asked Nola.

She took a sip of her water.

"All right. You've guessed something about me, and I'll tell you something that might interest you."

"I'm listening."

"My father was a diplomat. Or call him a diplomat. He lived in foreign countries—with his family, with my mother, me, my brother. He supplied information to people in Washington. I always wanted to emulate him. I attended a certain school in Monterey, California."

I thought I knew which one she meant. Tom and I had read and seen lots of vids about it. It was a spy school.

"So I was right that your interest in me goes beyond my charm," I said.

"If it does, Jeff, it doesn't mean that everything is a deception. Life is more complicated than that."

"You were going to tell me something about 'five-thirty.'"

She paused before responding. We sat there like two ordinary people in a tiny kitchen, at a cluttered table with a napkin holder and the hum of a nearby refrigerator, as if what we were talking about was not something incredible, unutterable.

"Let's say, Jeff, that certain people were keenly interested in information a certain scientist had in his possession at his death."

I nodded. "Let's say that."

"And let's say that they knew someone had borrowed some of that information and was seeking to convey it to a certain relative of the deceased scientist."

She saw that I was following her. I didn't have to say

anything. She resumed:

"And let's say that someone — perhaps a young person, a novice at his job was assigned to make sure that the information did not reach its intended destination. The Novice has essentially two targets: the messenger conveying the information, and the intended recipient. You understand, Jeff, that the Novice must keep everything he does a complete and undecipherable secret. And to do that, he refrains from using the real names of any of the people under his surveillance in any way. So instead, he invents a code. Person 'A' might become Apple. Person 'B' might become Banana."

"What's this got to do with 'five-thirty'?"

"I'm getting to that. Be patient."

As she was speaking, articulating this bizarre scheme of things, I could see in her eyebrows, in the movement of skin on the contours of her face — the lines around the mouth, a maturity I had not previously detected. I realized that she was older than I had been given to understand. How much older, I couldn't say — my age, I guessed, but even older than that in experience and sophistication. She'd been around the world, and in the corridors of hidden knowledge and power. I must have seemed like a bumpkin to her. I felt almost privileged to be taken this far into her confidence.

"OK," I said. "Apple and Banana."

"And let's say that while monitoring one of the targets, he finds that this person spends a lot of time with someone else, a friend, a neighbor. He's over at the neighbor's all the time. They go back and forth between their two addresses."

One beat. Then I got that she was talking about Tom and me. Two beats, and I understood why she had referred to the addresses: my address at the trailer park had been 530.

"Now suppose," Nola went on, "that the second target — the relative of the deceased scientist — has a different last name than the scientist had. And at this point, he really has no name. He is just Banana, and he has an address."

I jumped in.

"And say he has a friend, Tomato, with a similar address? Is that where you're going?"

She smiled, and I could see she was impressed. She was pleased, and that was almost thrilling to me—if I hadn't been so appalled.

"Jeff, agents are human beings. Especially novices. They can make mistakes. They can confuse things."

I felt almost sick, light-headed.

"God, Nola. Are you telling me that Banana becomes Tomato?"

"It's a simple mistake. Here's the Novice, out in the wilds of Ohio, dealing with what, to his people, have been—forgive me—non-entities. The targets are moving back and forth. One code becomes another. One address is similar to another, and gets confused with it. A mission is marked 'accomplished'—and even if a mistake is later uncovered, in a sense, the mission has been largely accomplished—although there may be a need for some follow-up."

"And you're the follow-up, Nola. You're the forbidden fruit, and you fall willingly into my lap, all in the line of duty."

She knew that the accusation would come, and she was ready for it.

"I didn't have to tell you any of that, Jeff. Some of the people I work for would not approve."

I smacked my fork on the table and jumped up, incredulous.

"What about the trial? Those people knew my name, what about that? How could they mistake me for Tom? I don't believe you!"

But she didn't have to say anymore. I sank to my knees on the hard linoleum floor of Nola's kitchen. I pounded my fists on the floor and wept, no longer able to control myself.

"My best friend is in prison for life! His wife and his child have abandoned him and gone to Canada! And it was all because of spooks like you who couldn't shoot straight and got the wrong man!"

57

Trust Your Feelings

It's amazing to remember it now, but it's very clear in my memory that I did not bolt, and that there was no further acrimony between Nola and myself that evening. Instead, I calmed down. I finished my dinner. (I wasn't about to let real food escape me, when I got the chance at it.) Nola then offered me some brand name Mary Jane, and we moved to the living room, sitting side by side on the sofa, flicking our ashes into a beige ceramic tray shaped like a horse's head with a wild mane, which in turn reposed on a smoked glass coffee table with curved iron legs. On the wall was a painting of a fox hunt: a gaggle of blue-blood Brits (probably) in red riding jackets, jodhpurs, black boots, helmets, and riding crops, atop spotted horses, trailing a pack of frenetic hounds away from a gated white stone mansion and into a deep green wood. No doubt, the scene was some whiff of Nola's high class background.

Then she reminded me of the offer I had made.

"You said you needed a contact, Jeff. It seems to me you've realized that you have one. If you could put this little matter behind you, I might have some time of my own."

"And how did you manage to just happen to work at that store where I just happened to be hired, Nola? Was it all part of the arrangements your people made? Tell me that, and then I'll make any agreement you like."

"It's not a difficult matter to line people up, if you know where they are. People are willing to cooperate with others who are—shall we say—very high up in the chain of command?"

I nodded, eyes closed. "OK ...OK."

I'd heard about this kind of stuff before. Like I said, Tom and I had read a lot of books and viewed plenty of vids about it. We knew our history, going back quite a ways. For example, have you ever heard of D.H. Byrd? He was the owner of the School Book Depository in Dallas, back when JFK was killed. He wasn't just some guy. He was friends with H.L. Hunt, and LBJ. If you don't know who those guys are, look it up. That's all I'm going to say.

"I think you know what my friends are looking for," Nola nudged.

"They can have it. They can have everything that belonged to my uncle, as far as vids and files. Only I'd like to have something non-controversial, just to remember him. He was my uncle, you know."

"Of course. Jeff, if you were to provide me with access to the recordings—all of them—I think I could locate what my friends are looking for fairly quickly. I could surgically remove it—and I'd leave everything else with you."

"That'd be OK."

"Should I come over then?"

"Yes. Tomorrow."

"After I get off work?"

"You're going to keep working?"

"For the time being."

"Let's say dinner. Maybe we can get some take-out and bring it back to my place."

"Sounds good."

"Does it? So you're willing to socialize with me, even though I'm giving you what you're after for free?"

"Jeff, I like you."

"Do you?"

"Can't you tell?"

"I thought I could."

"Trust your feelings. I'm a human being, Jeff, not just a robot carrying out a task."

"I should feel betrayed. I should feel used. But somehow,

Nola, I still like you too. Quite a lot, actually."

"Well then?"

"Well then."

Whatever you're thinking might have happened next did happen. I couldn't help it. I was twenty-six years old and bursting with vigor, and Nola was the best-looking and most intriguing woman I ever went out with.

58

As Obvious As a Mountain

According to Phyllis Camelot, nationally syndicated multimedia sexologist, a male orgasm brought about by means of mutually pleasing intercourse with a female (perhaps also with a male, for gay populations, although I don't recall her saying so) is followed by two or three minutes of a feeling of absolute, utter peace and well-being in the male experiencing the phenomenon. Depending on the youth and/or virility of the male in question, the period of absolute, utter peace and well-being is then followed by a gradually mounting return of the normal tensions as the wells of seminal fluid begin to recoup themselves for another assault on mortality.

Somewhere in there is where the cigarettes get smoked.

Myself, I don't smoke tobacco. But I was somewhere in that zone, post-intromission, when I rolled over and said to Nola, "What about my friend, Tom?"

She did not immediately respond. Then she simply repeated, with a period on the end, as if getting the words right: "What about your friend Tom."

"Yeah," I said. "I'm giving you what you want. I'm not causing anymore trouble, not like I ever meant to cause any. Tom is completely innocent, and he only got the hammer because he was confused with me, you say. So why can't we do something to get his situation straightened out?"

"I'm afraid that's impossible."

"Why? Why is it impossible? You mean, because then you'd have somebody on the hook for killing my uncle's secretary?"

"Mm-hm."

"God, somebody should pay reparations to that poor woman's family."

Nola leaned up on an elbow, looking at me with a stern, authority-tinged face.

"If you're going to cooperate with us, Jeff, one of the main requirements is silence. I can assure you that silence is in your best interest."

"Oh, Christ, Nola, are you threatening me?"

"No, I'm just pointing out what ought to be obvious, as obvious as a mountain or a skyscraper. There is nothing we can do about what has already happened. National security must always be protected, first and foremost. Everything gives way before that. Can you understand that, Jeff?"

"Why are you asking me if I understand? Do you think I'm stupid? I'm not stupid. I educate myself. I read. I watch educational vids. I have a geeky, vacuum-cleaner mind. I picked up on who you were pretty quickly, didn't I?"

"You sure did."

"Well, doesn't that tell you something about my native intelligence?"

"Yes, I think so."

"OK," I said. "So it's in my interest to keep my trap shut. So nothing can be done for Tom. So Ms. Pagano goes into the ground under an assumed name and maybe her family doesn't even know where she vanished to. I say nothing. I see, hear, and say no evil. I get it, OK?"

She studied me carefully with those Asiatic black eyes of hers, and I thought, I'd better keep a lid on it from now on, or I might end up having an accident or suddenly committing suicide.

But the very next day I went about deceiving Nola.

59

Uncle Preston's Speech

Shaker had told me that the operative vid was from 2025. My uncle had all his vids dated, and it wasn't a difficult matter to locate the mother lode. Before viewing it, I copied it onto a blank. The whole operation took maybe half an hour.

While I was doing this, I had the audios on. The government had sent federal troops to Texas, to quell the uprising. Despite the obvious disadvantages of owning real estate in the Gulf Coast, Uncle Sam did not want to give up his holdings there.

They even arrested some of the state legislators who had passed the secession bill.

We were making that big mistake of 1861 all over again it seemed.

As soon as I had the recording done, I cut off the audios and played back the copy of my uncle's 2025 contraband vid on my main mid-size screen. I'll do my best to describe what I saw on it, and to transcribe the speeches.

There's a mid-shot of a microphone in front of an empty, buttoned leather, swivel chair. This shot appears unsteady, but it stays in the same location, as if someone were recording it by means of a device on their clothing. There's a desk-like platform, gray, with dark wood inlays. You can hear the murmur of people in the room, not pictured. So this is the committee room. This unsteady but immobile shot and the murmuring last for several minutes. Every once in a while someone walks in front of the camera, usually a younger-looking person, well-dressed in male or female suit, sometimes holding a stack of papers or a book of some type. There's a name plaque in front of the empty chair. It says, "Chairman Welch."

Eventually the chairman comes in and sits in the chair. He looks about fifty years old, with very dark hair that looks almost wet, because it shines in the lights and is a bit wavy in the loose strands. The man has plenty of hair, despite not being young. His complexion is paler than the average, and his eyes are like charcoal bricks. He has an aquiline nose and what I'd call a haughty demeanor, which begins to show itself as he interacts with others.

He adjusts the microphone on the long desk in front of him, and you hear the amplified clunking of the contact.

"This is the Senate Subcommittee on Climate and Energy," he says, and he rattles off a number and announces the date. He declares the session open and asks a Ms. Harriman to take a roll call. Ms. Harriman calls out the names of several senators, and notes those present. A quorum is declared. All of this is done without us seeing anything but the face of Chairman Welch, which is occupied in looking at the senator's watch, or in listening to one of the senate pages who leans down into the view of the camera and whispers in the Chairman's ear. The rest of the time the Chairman's face stares impassively at nothing we can see. Finally, the roll ends, and the Chairman calls the meeting to order.

The Chairman announces that this is a continuation of a previous session, at which it was requested to report out a bill introduced by Senator Villiers of California. The purpose of today's session is to hear testimony from two expert witnesses before voting on the motion.

Now we hear someone's voice in the background, droning excitedly, but we can't make out the words. Welch is watching the speaker off camera, and the speaker evidently has no microphone.

"No, I'm sorry, you cannot take a video of these proceedings, sir," says the Chairman. "This is a closed hearing."

The objecting voice drones for a bit, the words unclear.

"I know what your First Amendment rights are, sir," says Chairman Welch. "You cannot stay. Please leave the room, sir."

Some of the words are now audible as the off-camera intruder says, "... people have a right to know ..."

"This is a closed session," says Chairman Welch. "Can the Sergeant-at-Arms ...?"

"... make a motion," someone says.

"Excuse me?" the Chairman says. "No, I'm not hearing motions now. We have to—"

"Why can't I make a motion to permit the hearings to be video-recorded?" says a male voice, obviously another Senator somewhere in the room.

"I'm not hearing motions. We're going to clear the room of unauthorized parties before we do anything. Will the Sergeant-at-Arms please—"

We hear a commotion, the intruder shouting. The picture now shifts jerkily, blurs, then we see a uniformed figure dragging a young man by the elbows out of the room. The young man is clinging to a video camera, but now, a second uniformed figure approaches and takes it from him. Then they all three exit through a pair of doors.

The camera returns to Chairman Welch.

"The first matter before us is the report commissioned by Dr. Preston Keys. Dr. Keys?"

"Thank you, Senator."

This is the voice of my uncle.

"Ladies and gentlemen, early in this century the voices of climate scientists began to come together on the issue of climate change with a degree of unanimity seldom seen. The percentage of quantified unanimity was often famously and correctly given as ninety-seven percent; tantamount to saying there was no opposition to the consensus. I cannot sufficiently emphasize that the consensus was clear and not seriously opposed by anyone in our field that greenhouse gases in the atmosphere, caused by man's burning of fossil fuels, had created, in effect, a heat-trapping blanket which was gradually increasing the surface temperature of the planet.

"As you know, the measurement of carbon dioxide particles in the atmosphere is done in terms of parts per million. It was noted, then as now, that this measurement had been estimated to be approximately 280 ppm or less for a period of some 650,000 years of Earth's history and prehistory, but that, since the advent of industrial civilization with its burning of coal and other fossil fuels, that figure had risen to 320 ppm, then 350 ppm, then 380 ppm, then over 400 ppm, and by the year 2015 was estimated to rise at least another 200 ppm and probably another 600 ppm to a high total of 1,000 ppm by the end of this century.

"What is currently in the atmosphere cannot now be subtracted from it. It will remain for another millennium. For at least the past twenty-five years, we have been observing the effects of having this carbon blanket in the atmosphere. The most recent example of such effects, which we are now calling the Deluge of '25, is demonstrating horrific human and financial impact on our lives and our civilization. I do not have to tell you what the financial impact has been. You have had it detailed for you in the many requests for federal emergency funds that have been inundating your meeting rooms, your desks, and your public fora, in a manner similar to the physical flooding of our cities by the moisture that now multiplies its weight and volume in an exponential manner, due to the physical properties of increased atmospheric temperature."

Uncle Preston could be a little long-winded at times. But I won't interrupt anymore:

"In addition to these impacts, we have seen the steady degradation of our environment, in the form of the acidification of our oceans, the melting of our polar ice sheets, the death of our coral reefs, on which so much of the marine-based food chain once relied, the unrelenting drought and wildfire in our western regions, the resulting lack of water in those regions, the loss of mountain snow melt, the contamination of our more abundant water here, in the regions east of the Mississippi, through algae blooms that result from increased fertilizer and pesticide runoff, due to greater rainfall totals which have the same cause as the present floods, the increasing size and intensity of our annual hurricanes, the loss of animal species including pollinators on which we once relied for the abundance of our supply of fresh fruit, the impacts on fish and wildlife, the steadily increasing and increasingly unmanageable costs of insurance, with their corresponding impacts on investment, and thus on our economy.

"Those last matters I am aware of, although they are not in my sphere of study. I am a scientist. I study the impact of climate on our environment. I speak of my sphere of study when I speak of the environmental impact that I have outlined, but I add the financial impact at the end, because I know you are aware of how those costs impact the whole arena of investment and finance, and that you are sensitive to such issues, for you have many influential people from the world of business and finance who come before you in this body, often. And many of you come to this body from that world. You are lawyers, bankers, businessmen and women.

"Ladies and gentlemen, we are not merely at a crossroads. We are in the path of a road already taken. But we are on a final stage of it that we need not take. In the first decade or so of this century, it was becoming more and more apparent, to those who follow such matters, that the Middle Eastern oil on which our economy was based was reaching the end of its viability as a plentiful source of energy. We were running out, and we knew it. That was the period which you may know as Peak Oil.

"It would have been a great boon to us had we simply run out of oil. Our economy would have collapsed, or at least been severely affected, and we would have bemoaned our condition, but it would have been a boon in disguise, for it would have meant that the burning of our fossil fuel supply would have greatly diminished. We would have continued to burn coal, but the level of CO2 parts per million in the atmosphere would possibly have leveled off due to much lower levels of the use of fossil fuel based energy.

"Instead, we discovered the process of horizontal fracturing, or 'fracking', as it came to be known. And in our zeal to commence a new and excitingly domestic energy boom, similar to the early days of oil exploration when we had wells pumping oil in Texas, we rushed to invest in the new fracking boom. It meant jobs, it meant energy independence from OPEC, it meant cheap gasoline, if we could create it out of natural gas obtained through the fracking wells. And there were businesses devoted to this last enterprise, as you well know, and they have become the rising new companies of our day, and very profitable for their shareholders, I have no doubt.

"This event was the undoing of man as a species, and it meant the end of the planet as we have previously known it. We have now begun to see some of the harmful, as well as the beneficial, effects of fracking. They include contaminated water wells, increased and unacceptable levels of arsenic, barium, DEHP, glycol compounds, manganese, phenol, methane and sodium in our water. DEHP, in case you're wondering, is a phthalate—p-h-t-h-a-l-a-t-e—a substance occurring in many household items, which leaches into human blood and damages the endocrine system and can cause reproductive system cancers and defects. The fracking wells, which were supposed to provide endless domestic energy sources, instead send energy to China, while at home their proximity destroys property values as well as health, and in order to be offered a halfway acceptable buy-out by the well owners, the sellers are forced to sign covenants not to sue and not to speak or make known what has happened to them.

"But I do not speak to you as an advocate for public health. I speak to you as an advocate for the health of the planet, which is a sine qua non of our existence. I am telling you flatly that we cannot go on burning fossil fuels without any controls on our CO2 emissions. It may already be too late, but it is my hope that some form of adaptability may be possible if we can only cause the emissions to level off to a stable amount. Continued increase is simply fatal to our world.

"I am not here to tell you that we can solve all energy need issues with a combination of renewable sources. Others will tell you that. They will say the technology is here, we can do it. I hope they are right. What I am saying is that, regardless of whether or not they are right, we will end life on this planet as we know it by continuing on our present path.

"I do not say this with a lack of awareness of what this means. I am well aware that major food sources in our country have, in the past, been reliant on high-technology agribusiness, and that these businesses have been largely funded by the oil industry. I am aware that shifting to a form of agrarianism based on an Amish model will result in great decreases in the food supply, and that this will cause hardship, even perhaps a degree of famine in our country. However, I would note that this process is already underway. I have been told, for example, that there will soon be a bill before you, and before the House of Representatives, that calls for national food and water rationing.

"Early in this century, even in the second decade of it, there continued to be many people, even in this august body, who persisted in refusing to see the downside of their position. They denied the effects of climate change. They said it 'required greater consensus, more study', that it was a 'complicated issue', that, in fact, the changes that were then being observed in weather patterns were not necessarily attributable to greenhouse gases, but were part of naturally-caused patterns, or at least could not simply be assumed to be otherwise. Some said the effects were not expected to be severe as this century progressed.

"I do not make the mistake these people made, of ignoring or downplaying the arguments in opposition. I do not say that those who disagree with me have no argument to make. They do have an argument to make. They are right that there will be severe economic impacts if we change the basis of our economy from a fossil fuel based economy to another one. But I am saying there is no choice, if we want to continue to survive as a species.

"Mr. Chairman, I express today my support for the bill that is being put forward by Mr. Villiers of California.

"I thank you for your time and your patience."

60

Moonbust Petroleum & Shale

Throughout this speech of my uncle's, the camera remained unsteadily focused on the Chairman, Senator Welch. As I was listening to Uncle Preston's remarkably controlled presentation (for despite the provocative words, his voice sounded rational, well-modulated and determined) I had to endure the Chairman's facial expressions, which varied from supercilious, to annoyed, to disdainful. Toward the end he was rolling his eyes with undisguised impatience. I wondered how someone in his position could be so disrespectful of my uncle, a man devoted to science.

The hearing was not over at that point, however.

There was another speaker. The Chairman introduced him as Rowan Chenevey, Chair and Chief Executive Officer of Moonbust Petroleum & Shale Investments, Inc. As Chenevey began to speak, the camera shifted shakily and blurrily to one side, and we could see the speaker.

This was when I realized that the recording was being secretly made by my uncle, probably by means of a tie-clasp camera, which would be aimed outward. So that was why we saw only Chairman Welch during my uncle's speech, because my uncle would have been facing him and addressing him. Now, however, he could turn and face the other speaker.

The secret nature of this recording was also obvious by the fact that another would-be reporter had been kicked out of the hearing earlier, and the fact that this recording of my uncle's had become of high national security importance to our government.

Rowan Chenevey had a large head, round on top, with thin, graying hair. He had high cheekbones, a longish face, and a large lower jaw with lots of teeth. He had a loud, emphatic, medium baritone voice, like someone accustomed to being listened to and

Clarke W. Owens

obeyed. He wore a shiny black suit and a dark blue tie over a white shirt. Next to him, on his desk, was a small glass of water. Here was his speech:

"Thank you, Mr. Chairman. I am delighted to address you today. I see many familiar faces in this room, and it's an honor to be here before you to dispel some of the misconceptions that continue to plague our industry and to hamper the continued prosperity of the American people.

"I have listened carefully to your previous speaker, and have heard the same clichés and distortions of the truth that we heard twenty years ago, in the early years of the boom in our enterprise with horizontal hydraulic fracturing. I say the boom because, as you may know, hydraulic fracturing is a process that has been with us since the 1940s. With advances in horizontal access technology, we were able to achieve the nearly miraculous and highly productive results which have saved the American economy from what otherwise would have been an ongoing, perhaps never-ending depression. Instead of that bleak result, we have provided jobs and incomes for our people, an inexpensive supply of domestic energy, and hope for the future.

"The claims about dangers to public health and water supply contamination are false. They are bugbears that have been with us since before we even got the industry fully underway. The idea that we are pumping huge quantities of poisonous chemicals into the ground is ludicrous. 99.2 percent of the fluid used in the process is water. The chemicals that are used actually help protect public health in the long run, by protecting the integrity of the product. For example, corrosion inhibitors and biocides. Without these types of chemicals, you could eventually have product contamination, which would not be healthy and would also be economically counterproductive.

"Ground water contamination from a frack well is impossible. The tops of the fractures are thousands of feet below the aquifers. They are not even close, and therefore, contamination does not occur. The steel casings used are impregnable and placed at depths of one to four thousand feet. The cement used for the annulus does not allow for movement between the strings and the wellbore. The surface areas are ecologically well-integrated and are no longer based on the old models requiring vast spaces occupied by multiple well heads. Instead, we consolidate the well heads into a single pad at most sites. The surrounding area is untouched. The horizontal outreach occurs thousands of feet below the surface. Gas recovery does not occur above six thousand feet in most instances, and often well below that.

"What people like the previous speaker want is to stop progress. They want to stop growth. You heard him: we should go back to Amish farming methods. Well, you and I know that if we did that, we would all soon starve. There are four hundred million people in this country. You can't feed that many people with Amish farms. I guess he wants us to stop using tractors because they run on gasoline.

"You heard the previous speaker admit that declaring a moratorium on the oil and gas production in this country would bring the economy to a grinding halt. We would be thrown into another Great Depression. Only it would be worse than the Depression of the 1930s, because there would be no incentive to restart the gears of economic progress. Economic progress is the enemy for people like the previous speaker. People like you and me, who believe in economic progress, are considered inhuman, as if we did not care about clean air and water. Well, I have children and grandchildren, and I do care about clean air and water. But I know something about the industry the previous speaker criticizes. He doesn't.

"As far as greenhouse gases, we have been hearing that gloom-and-doom prophecy now for at least twenty-five years. 'The world's going to heat up.' 'The polar bears are dying. They don't have any ice floes anymore.' Well, I feel real sorry for the polar bear. But I care more about the future of my children and grandchildren than I do about the polar bear, excuse me. And I don't want my grandchildren to grow up in a third-world country with starvation and poverty. I want them to grow up in the United States of America, the greatest country in human history, where the future is always bright and where there is always a promise of affluence for the future.

"Ever since human history began, people have been proclaiming the End of the World. It hasn't come true, has it? Can you still breathe the air when you go outside? Have you had anyone in your family die because they drank the water from their kitchen tap? I sincerely hope not.

"The fact is, the predictions of a dead planet due to carbon dioxide particulates in the air is a radical, anti-capitalist hoax, perpetrated on you by people who hate freedom and progress.

"If you pass the bill drafted by the Senator from California, you will send a shock wave through my industry that will bring down the pillars of prosperity in this country for a long, long time to come. Investment will cease. And that means investment in government as well, I hope you realize. There are many of us who invest in the careers of well-meaning government officials such as yourselves, as you well know. And I guarantee that if you cut off the legs of industry, industry will not see fit to continue to support a system which no longer believes in free markets and the American Dream.

"Thank you."

61

Last Gasp

The next part of the recording was devoted to debate by the senators of the bill proposed by Mr. Villiers of California. Those who spoke in favor of the bill said that it was the most important vote any of the senators present would ever take and that if they killed the bill, they were driving a stake through the heart of ongoing life on Planet Earth. Those opposed to the bill said that it was 'shrill' and 'alarmist', and that it would not be wise to bring the economy of the most important country on Earth to a complete standstill for a benefit that was, at most, speculative. There were many more of the latter type of speeches than the former.

When the speeches ended, a voice vote was taken, and Chairman Welch declared that the Nays had it. A roll call vote was then demanded by someone on the floor. That was done, and the number was quite lopsided.

The bill had been defeated and would not be reported out to the larger committee, and thus, not to the floor of the Senate.

It was dead.

I suspected that what I had witnessed on this vid had been, some twenty-five years ago, the last gasp of a chance that any of us had to avoid the kind of world we were now living in.

I made a copy of the vid and took it to my bank where I placed it in a safe deposit box, along with another item. I kept the original with all the other media left by my uncle, together with all his indexing, in a box ready for pickup by Nola Sheeka—or whatever her real name was—that evening.

62

The Next Stage Down of Helpless

Although Nola had suggested she would not take everything left to me by my uncle, she did, in fact, take it all. She said it would be easier to send items back to me than to take a chance on allowing any 'unauthorized' material to remain out. I said she could do as she wished. I simply wanted to be free of any controversy, and not to be on anyone's 'enemies' list. I said I just wanted to live my life and be left alone anonymous and insignificant as I was.

Within a few days, Nola had left her job at the market and had disappeared back into the corridors of secret power from whence she had come. The whole episode with her seemed like a strange dream, but I began to feel slightly less paranoid, worrying now only about the remote possibility that I might have been seen making the safety box deposit at the bank.

It was later that same week. The rains were insistent and the skies remained dark. There was flooding in several parts of the state, and in parts of our county. I had taken to walking to and from work because the roads were too inundated with water to ride my motorized bike. On my route home, I passed some commercial buildings in the downtown area located near my neighborhood. At night, these buildings were deserted.

As I passed a gap where an alley separated two of these buildings, someone leaped on me from behind. I felt a gloved hand wrap itself around my mouth and an arm tighten around my neck. At my right ear, I felt some type of hot cloth or cottony material, as if the person who had jumped me were wearing a ski mask. I felt moist, hot breath there, and a low voice said, "OK, Claymarker!" I felt a shock of fear go through my body.

In the instant in which the assault occurred, it flashed through my mind how easily my assailant could have knifed or shot me. At the same time, there was a curious clumsiness in the way he had grabbed hold of me, and—if this is possible—I sensed a certain feebleness or weakness in the arm lock. The arms themselves felt thin and light and the body behind me also had a light, slender feel to it. This was the impression I had and I thought it strange, but it was the impression of an instant.

In the next instant, I felt myself falling backward, for my shoes had lost their footing on the slippery-wet pavement. I felt some pebbles sliding underneath the soles and heels, heard their gravelly churning and spitting as I began to fall, and as I realized I could not stop the fall. The assailant clung to me almost like someone riding piggyback and we both fell backward into the drizzly, pebble-strewn alley.

It was his own fault that I landed on top of him. The breath whooshed out of his lungs, and I heard a nasty whack as his head hit the side of one of the buildings forming the alley. As I struggled out of his now dazed arm lock, I realized he was incapacitated by the blow. If he wasn't killed by it, it was due to the fact that he did have a ski mask on, which cushioned some of the impact. It was a navy blue ski mask, with orange horizontal zigzags.

I was keenly cognizant of the fact that he had used my name and I wanted to know exactly who he was, thinking it must be one of the spooks who had made their presence felt in my life for several months now. I half-expected to see that kid who had held up Tom for the water card when I reached down and peeled the ski mask off, from the chin upward, with no resistance from him whatever.

"Joeg!" I gasped.

For Joeg Hibben it was, blinking and semi-conscious, working his jaw as if to clear a few loose marbles in his benighted head. I could see he had no weapon on him and that he was virtually helpless at this point.

I unleashed my fury on him.

"You idiot! What are you doing jumping me like that? I could have you put in jail, you moron."

All the fight was out of him. He addressed me as if we'd been pals, fellow co-workers at the store with no animosity between us.

"Don't do it, Jeff. I'll lose my job. Don't tell on me, Jeff!"

"This is about Nola, isn't it?" I demanded. I knew, of course, that it was, even without asking.

"I loved her," he said, rubbing the back of his head with one gloved hand. "She's gone now. She left because of you."

"No, she didn't, you dumb shit. Nothing she did was because of me, believe me."

He had been staring straight ahead and blinking, but now he looked up at me with a kind of pathetic, hopeful, inquiring gaze.

"Really?"

"Really. She had a big important job she had to go to, back East. It didn't involve me at all. I'm nothing. I'm a stock room guy in a grocery store, like you. Hey, how's your head?"

"I'm OK," he said.

"Well, you'd better go to the E.R. and have it looked at. That crack didn't sound good when you hit the wall."

"No, no, I'm OK," he repeated. "Don't report me to the police, Jeff. I'm sorry I jumped you. She's really gone for good, is she?"

"She's got bigger fish to fry. I won't turn you in, if you promise not to pull a stunt like that again."

The backs of my trouser legs were soaked through, as was part of my jacket. I helped Joeg to his feet.

"I'm sorry," he repeated. "I lost it. It won't happen again."

I handed his ski mask to him and said to forget it. He turned and slunk away, embarrassed and defeated.

My anger dissipated in the cold drizzle. I couldn't hate or fear somebody as pathetic as that. I knew he was the next stage down of helpless from me. I went home to clean up.

63

Serendipity Again

I kept a low profile for the rest of October and into November. The political season was getting underway because 2052 would be an election year. I found it hard to listen to Wesley Wright anymore, because every time he would talk about how you should never question America or the need for its endless wars, I couldn't help thinking of the spooks that had invaded my life and hurt me and those close to me so badly. I shut off the audios when Wright came on. No more Patriot Network for me. The spooks had lost one of their own little lambs when they lost me. I had no doubt that they controlled those air waves, anyway.

So here came the Thanksgiving holiday, and I sure didn't look forward to it. My mother was dead. My best friend was in prison. His family was somewhere in Canada. My girlfriend had broken up with me. My little fling with Nola had turned out to be a dance with the devil, and now I was all alone. Alone is what you don't want to be during the holidays. They say more people kill themselves at that time of year.

And on top of that, I had a birthday coming up, a couple days after Thanksgiving. I would be twenty-seven.

Fortunately, I still had my little job at the market. Also, Cyril Shaker's office had refunded a portion of my five thousand dollar deposit which hadn't been used. And then my neighbor, Grady Lurie, invited me to spend Thanksgiving dinner with him and his family. He said they'd be serving real food. This was a ritual that most families who could afford it observed. I offered to contribute and he said, just bring something from the market, if I wanted and could afford it. I had a discount there.

I guess it was a week or so before Thanksgiving that I finally got another letter from Tom. He said that Irene and Tommy had been his soul. He said that life without them was like a plant or a flower trying to live without rainwater or light from the sun. He said that was exactly what he felt like in prison. He said he hoped I was OK, and would I write to him once in a while.

I started a letter to him but his issues were so big that everything I said sounded phony, insincere. It wasn't insincere, but it was like I couldn't address anything in his letter, because his letter was one big expression of pain, and when someone is suffering like that, you know damn well that nothing you can say can make a difference.

So then I thought, I'll just keep it small. I'll tell him what I'm doing, tell him what's new around here, and maybe that will help console him in some minor way—just seeing that there's another world out there, besides the hell he lives in.

Then along came Serendipity again. It was the next day after the letter from Tom that I got a skype call from somebody at the University of Akron. It was a person who looked just about my age, a woman, sitting in an office of some kind, with dark green drapes behind her. She looked Hispanic. She was dressed smartly, with a kind of bow tie or string tie on a ruffled white blouse, dark skirt and matching vest. There was a young man next to her, but half of him was off camera.

"Jeff, how do you do, I'm Juanita Lumens from the University of Akron School of Law Innocence Project. I'm a third-year law student. This is my associate, Warner Phillips."

The half-out-of-the-picture guy nodded and said, "How do you do."

What I could see of him looked even younger than the woman.

"We received information from a gentleman by the name of Cyril Shaker about your friend's case," said Ms. Lumens.

"Cyril Shaker is dead," I blurted.

"Yes, we know. But he spoke with us before he died. We think we might be able to help you. Are you able to come to Akron so we can talk to you? There's no charge."

She could help me at no charge!

"You bet I can," I said. "When?"

"Next week, perhaps?"

"I have Mondays off. What if I came in Monday?"

She said that would be fine. We set a time late in the day, so I'd have plenty of time to get up there. I'd have to find a bus or a cab, or someone to take me. I didn't think my little motorized bike was up to it.

I decided to hold off completing the letter to Tom until I could get up to Akron on Monday. I figured I might have some news for him then.

64

Physically Speaking

I made an arrangement with a local cab driver to take me up to Akron and wait, and bring me back, for a fixed fee. He was happy to do it, and I used funds from my refund from Shaker.

The law school was on a corner of the urban campus, with the downtown only a block away to the west. There was a neat collection of parking lots near the building, which was clean and pleasant looking, next to a stoplight near a performing arts auditorium. I entered the building on the south side, through a glass door, as I'd been instructed, and walked down a corridor and turned right, then through a wooden door marked "Appellate Clinic."

Inside, I told a secretary who I was, and that I had an appointment with Juanita Lumens. She buzzed Ms. Lumens and I was sent through another door into an office, where I encountered the same two young people I'd seen on the skype call. They both smiled and urged me to sit down.

I recognized the same green curtains I'd seen in the skype, bunched together like bars at a window behind Ms. Lumens's desk. There were bookcases full of, I guess, law books, and there were lamps on dark wood furniture, and the room had a deep smell of polished wood. The chairs were well-cushioned and comfortable and I had a protected feeling, as if the spooks couldn't get me here, surrounded by the law. That was probably an illusion, though.

After we re-introduced ourselves, Ms. Lumens said, "We're just starting an Innocence Project here at UA. Warner and I are both third-year law students. We work under a faculty member who's not here today, but you'll get to meet him eventually—Professor LaSevere. We learned some facts about your case from

Mr. Shaker. There was no intent to violate your confidentiality, but he was aware that you'd need representation at some point and he'd heard of our project, so—"

"I'm not worried about it," I said. "What did Mr. Shaker tell you? You know he was murdered, don't you?"

They both hesitated on an intake of breath, exchanged a glance with each other, then exhaled.

"We had understood he committed suicide," said Lumens.

"No, it wasn't suicide. But tell me what you know, and then maybe I'll have more to say."

As they talked, I realized they knew everything that Shaker knew about the water card set-up. They didn't know about the Senate subcommittee meeting because Shaker had only uncovered that just before he'd been snuffed out.

I told them about it.

And I told them about Nola.

Warner Phillips furrowed his otherwise smooth brow.

"Are you saying that this woman, Nola, now has all the recordings your uncle left you?"

"She's got them all, but I kept copies of two," I said.

And I fished out an itty-bitty flash drive from my pocket. On it were the recordings of Uncle Preston's reference to the clandestine water card program and the complete Senate subcommittee hearing of 2025. These were the two 'items' I had placed into the safe deposit box at my bank. They were really one item, physically speaking.

illustriousancestors.com

After I explained to them what was on the recordings, they both developed very serious expressions on their young-like-me faces. I realized I'd dumped more on their shoulders than they'd bargained for.

"You understand," said Warner Phillips, "that our main focus will be on attacking your friend's conviction."

"Yes, I understand."

"We don't know," said Juanita Lumens, "if what we have will be enough to get the conviction entirely thrown out, but we think we might be able at least to get a new trial ordered."

"Mr. Shaker said he was going to be our main witness," I offered, "but that's obviously not going to happen."

The two of them exchanged a sly little glance.

"Do you think your friend Tom would be willing for us to take his case?" Lumens wanted to know. "We're volunteers, so there's no charge to him."

"He'd jump at the chance."

This made them both grin.

Phillips said, "We have his address and have already sent him a contract. We'd appreciate it if you'd contact him and put in a good word for us."

"I'll put in several good words. I just recently got a letter from him. This will give me some good news to send him."

We were all getting along swimmingly.

"There is one suggestion we have for you," said Lumens, "as a preliminary matter. This is something that will absolutely strengthen Tom's case."

"What's that?"

"We think you should donate a copy of this flash drive to an acquaintance of ours by the name of George Natale. Mr. Natale owns an online genealogy research company. What's it called, Warner?"

"Illustrious Ancestors, Inc. Or illustriousancestors.com, I guess."

"Yeah, Illustrious Ancestors. Anyway, he collects genealogical information and family records. He has a huge archive of such records. We strongly, strongly suggest that you authorize us to make copies to donate to his archive. Are you willing to do that?"

I sensed from the way Lumens was asking me that there was a strategy behind it, and that it was going to help Tom in some way. So I said yes. They gave me a piece of paper to sign, and I signed it.

Then we all shook hands, and after some more talk which I can't now recall, the interview ended with a caution that I should communicate only with them and with Tom. I would be introduced to Professor LaSevere at a later date, they said.

I went back out to the parking lot to catch my cab. It was late afternoon and the skies were gray. A wind was kicking up. I felt a degree of uneasiness from being matched up against so much power, but at least I felt now as if we had a fighting chance. And since we were fighting for justice, it was worth it.

66

You Have to Say Yes to Something Nutty

I endured the holidays by not observing them. I pretended that each day was just another Thursday or another Friday, etc., and that worked remarkably well at keeping me from being depressed. No, let me take that back. I did go to Grady Lurie's for Thanksgiving, and it was not so bad, except that Grady's wife, Cassandra, was Christian, so she had to deliver a prayer before the meal. I'm OK with that. I keep my head down and my thoughts to myself in those situations.

I was raised a kind of Christian myself. At least, my mother used to tell me we were Christians. We never went to church, except sometimes on Christmas or Easter. When I reached my teen years, Mom acted surprised when I expressed skepticism about the whole thing. We had a conversation about it in the kitchen one night after dinner.

"The thing about Christians," I said, "is they explain everything so that God always gets the credit for good things and never gets the blame for bad things. If their best friend gets cancer, they say, 'This is a trial, but God will get us through.' Then they pray and pray that their best friend gets well. If the cancer goes into remission, they say God answered their prayers. If their friend dies, they say, 'God took my friend because He wanted her with Him. She's in a better place now. If a tree falls on your house, but you escape getting killed, they say, 'God was looking out for me.' But they never ask why God let the tree fall on the house."

"Well, don't you think God has a plan for us?" Mom asked.

"That's what Christians always say. Not just a plan for 'us', but also a plan for 'me'. Everything is about them. Their eternal life, their well-being, their blessings. They don't care at all about some cave man who lived in prehistory. He can die and be dead, and it's no problem. But they have to live forever, with their own memories, throughout all eternity. That seems self-centered and nuts to me."

"Well, don't you believe in a Higher Being, Jeff?"

"I don't know. I think dolphins and turtles and elephants were probably higher beings before they went extinct. Maybe there's higher beings living on different planets. How would we know?"

"Well, don't you believe, Jeff, that Nature is an intricate design that must have intelligence behind it?"

"Why must it have intelligence behind it? And why must it be a man's intelligence? And if there's an intelligence behind it, how come the Christians are always saying God only cares about people, and only His people? If I'd created every spider, butterfly, and mongoose on the planet, I'd sure as heck care about all of them as much as I'd care about humankind."

"Jeff, do you think that the universe is just an accident?"

(You could tell from her questions that she was getting more and more worried about me.)

"What does it matter what I think, Mom? I'm just a tiny little moment in the universe. If you look at evolution, it looks like it starts with something less complex and evolves into something more complex. That suggests that we started with something less complex, not some great Mind."

"Well, Jeff, don't you believe in something larger than yourself?"

"Mom, everything is larger than myself except ants and amoebas. If I were larger than everything, I'd know something about it. I don't. This is humility here, not pride."

"So you do believe in something larger and—and Higher?"

(I mean, you could hear these capital letters.)

"Oh, heck, OK, Mom. Something larger and higher. OK, I'll go with that."

I had to give her something, don't you see? I had to say yes to something, or she'd have kept going until I did, and we would have been there all night. That's how it is with people: you have to say yes to something nutty, or they won't let you off the hook.

67

Self-Deception as Art

Sometime in December a new state law went into effect, requiring there to be no unoccupied bedrooms in houses. If you had an unoccupied bedroom, you had to accept an American refugee into it. The refugee was required to pay rent, or if he/she couldn't pay rent, you were required to be paid by the State for the refugee's rent obligation.

There was a legal definition of 'unoccupied bedroom', which I became familiar with because I was renting a three-bedroom house. A bedroom was a room in a house which had been designed or intended primarily for use as a bedroom, regardless of how it was currently being used, except that you were entitled to claim one such room as a den, library, or study.

The refugee had to be an American citizen who had left his former place of residence to avoid climatological effects such as drought, water contamination, destruction of one's home due to acts of God (wildfire, flood), or toxic environment. The rules went on and on and there were already several lawsuits cropping up, claiming that the law was unconstitutional.

I had one extra bedroom, but I got together with my landlord and we declared it a den/study, so I avoided having to share the house with a refugee. I think I mentioned before that I had refugees living next door. The refugees were everywhere. Half the time you didn't know they were refugees, because they were Americans.

I considered myself lucky to be living in one of the areas where the refugees were coming to, as opposed to the areas they were running from. Things can always be worse.

The other thing that happened in December was that the people from the Akron Innocence Project filed a petition in the court to grant Tom a new trial. I wished I knew where Irene and Tommy had gone, because I wanted to tell them about it. I wanted them to know Tom was innocent, and we were going to have a shot at proving it.

Then it occurred to me: why can't I find Irene? I started doing internet searches. If there was a public record somewhere, it would have to turn up. There would be ages given and people's names who were associated with the searched for person, so you could usually figure out if you had the right one or not.

I came up with a handful of Irene Glendinnings, but the only one that was obviously the right one had her still living here in Wooster. I knew she wasn't here. She was up in Canada or someplace, but she hadn't gotten arrested and hadn't bought a house or something like that that would create a public record that could be traced. Naturally, I didn't write to Tom about doing this, as long as I was coming up empty. That would only make him more anxious.

I called the Akron people every so often to check on how the petition was going and they always said the same thing. "We're waiting for the judge to rule."

So we just had to wait.

So much of life is waiting that you have to learn to make it an art form. I decided to do this. The first thing I did was to stop calling Akron. Then I pretended like I forgot all about the petition. I read books. I drew cartoons. I listened to music. I watched movies. If Grady Lurie asked me how the petition was going, I'd say, "What petition? Oh, you mean the thing with Tom." You know, self-deception as art.

It wasn't until January that we got a decision.

The judge granted Tom's request for another trial based on newly-discovered evidence.

68

El Refugio

The male half of the Innocence Project from Akron called me a week or so after the decision permitting a new trial. He wanted to talk to me in person. I knew it was not about the new trial, because that decision had been appealed by the State of Ohio and now we were waiting for the appeals court to rule. I told Warner Phillips it was really hard for me to get to Akron and he said he would come to Wooster if I could give him a date with an hour or two to talk. I gave him a Monday, my day off.

Warner—he told me to call him by his first name—met me at a local Mexican restaurant called El Refugio, which had real food. He bought me lunch. The place was located on Burbank Road, which was a massively commercial area sprawling northward to the county line.

We sat in a booth with garish Mexican, plastic, wall decorations everywhere—cactuses and burros and sombreros and senoritas with frilly-sleeved white blouses and brightly colored flowing dresses, logos for Mexican beers, things like that. There were arched stucco doorways, painted pink and orange and fake palm trees in big pots near the arches.

We ordered a couple of Mexican beers to start and Warner began speaking in a low, nervous voice, as if he were afraid someone might overhear us. He was dressed nattily in slacks and a pullover, sleeveless sweater, long-sleeved sport shirt, blue with white pinstripes. He had a florid face with short-cropped brown hair and the effect was to make him seem younger than he was, like a high school track star who'd just run a long race.

The weather outside was in the forties.

"I want to talk to you about your other recording. Not the one we're using for Tom's case."

"You mean the Senate subcommittee hearing in 2025?"

He nodded. "I've had some differences of opinion with Juanita about it. She says we should just forget it, it has nothing to do with our case. But I think—" He hesitated, glanced around him, then lowered his voice. "I think it's a bombshell."

I smiled. "That's what I thought."

"I mean, I don't think people realized what took place."

"How would people know?" I said. "They didn't allow cameras."

"Well, listen, Jeff. You had said that Mr. Shaker was murdered. You know the coroner didn't find that."

I didn't say anything, but I guess I must have grimaced.

"And when I talked about it with Juanita, she just dismisses that. She says the coroner made a finding and that's that. But you said he was murdered, and—well, I wondered what made you say that."

"You really want to know?"

"Yes."

"I was on the phone with him when he died. He was trying to tell me how I'd been confused with Tom by the spooks. He was in the middle of telling me about it when somebody shot him."

"Was it a skype call?"

"I had no visuals on his last call, only sound. But I heard voices in the room. I heard somebody cussing. Then I heard a pop and then the line went dead. You can't tell me somebody decides to commit suicide while in the middle of relaying an urgent warning to someone else. And you can't tell me I didn't hear the other voices."

"And—and your mother, Jeff? It was a burglary, isn't that correct?"

"Sure, it was a burglary, but they left valuables lying around. They took only media. They were looking for my uncle's stuff. Look, it doesn't matter to me if you believe me."

Warner's eyes had closed briefly while I was telling this. He nodded slightly, then he opened his eyes and ran his tongue nervously over his lips.

"Jeff, I believe you. But it's only me that does, OK? Juanita is not on board with it and I haven't even brought it up with Mr. LaSevere. I'm not going to go to them on this. But I want to talk with you about—well, about what might be done."

"What do you mean?"

"It's history now, Jeff, but it's big. It's a big, big story that has never come out into the open. The U.S. Congress sank the last chance we might have had to stave off the wreck that our environment has become, you see?"

"Oh, yes, I see," I said.

I did see that, if you'll recall. And, as a matter of fact, I had a half dozen itty-bitty flash drive copies of the 2025 subcommittee hearing still in my possession. But I didn't tell Warner Phillips that just yet.

"Well, I think—I mean, I want to know what you think should be done, Jeff. Because you own this information. But it seems like something that it might be in the public interest to go public with, at some point. If you see what I mean."

"'Go public with'? You mean, like what? Send a copy to the New York Times?"

He looked surprised, like I had guessed his exact thoughts.

Right then the waitress came to take our food orders. When she left, Warner picked right up.

"That's what I had in mind. The Times or the Washington Post, some big newspaper with tentacles into all the major media."

"Of course, the major media are controlled by National Security."

The way Warner looked at me now, I felt like I was an adult talking to a boy. I could tell he'd heard this kind of accusation before, but had not been accustomed to taking it seriously, until now.

"Would you be—concerned about that?" he said.

"Oh, yes, I'd be concerned. Wouldn't you, if your mother and your hired private eye were both dead, and your best friend was in prison on a frame job after being confused with you? Am I excessively paranoid, or does this make any sense to you?"

"Should we just forget about it?"

So here I was. Here was this nice, privileged, young man, soon to be a lawyer, sitting on a powder keg of historic proportions, looking to little old Me, waiting for instructions. And I could take the powder keg and bury it, or I could detonate it.

"Warner," I said, "what I think we should do right now is focus on Tom's case. We should try to get Tom out of prison and his name cleared. If and when that happens, I'll give you a call. I'll hang onto your number at all times, and if it changes, you should let me know. And when I give you that call, you should send your copy of the 2025 hearing to the biggest-name editor you can find at the Times. But not before I give you that call."

He nodded. His eyes were alight now. I could see we were on the same page.

"I understand."

"The biggest-name editor you can find. Or better yet, the biggest-name investigative reporter."

"Somebody fearless."

"That's it. But not before the call."

"Not before the call."

And that was because of the danger, obviously. From now until the time Tom was cleared was the most dangerous window. It was my fault Tom was in prison, so it was up to me to make sure he got out. I didn't want to jeopardize that mission. But once it was accomplished, I was willing to take my chances.

69

Not Trying to Hang Your Puppies

In January, I got laid off from my job at the market.

"I had hoped," said sad Mr. Barrington, wearing a red apron with big pockets and wiping his hands on a rag, "that the economy might pick up and hold on, but it seems things just keep getting worse and worse."

Six of us got laid off, including Joeg Hibben. I think Joeg and I each consoled himself with the fact that the other got the axe as well. I felt like Lee Harvey Oswald, being pushed around like a marker on a chess board—someone being used until I'd served my purpose, then disposed of. You remember I told you that Nola basically admitted her people had fixed me up at the market to begin with.

So it was back to the temp agency and a job at the brush factory.

One night, after work—this was toward the end of the month—Grady and Cassandra Lurie dropped in to visit. Cassandra had the baby with her, and she carried a big square knitted basket with infant supplies in it. She took out a bottle and fed the baby while Grady told me some breathless news in a strangely cheerful tone.

"Guess who we saw at the Big House tonight?"

The Big House was a local restaurant in an old building that had once been used as a jail.

"Who?"

"Kareena."

"Oh?"

I didn't want to act too interested, but life was lonely these days, so I waited to see what was coming next—and it wasn't good.

"And guess who she was with?"

"I can't guess."

"Gus Maxton."

I blinked. "Who's Gus Maxton?"

Grady laughed—not exactly an amused laugh. It was more like, it figures you wouldn't know.

"He's a heroin dealer, dude. Big time."

Grady sometimes knew too much about such things. I'd never been as close to him as I had been to Tom. But whatever his connections were, he was probably on the fringes. Or maybe it was only that he knew people who knew people. Hearing this news sank my boat, though. I knew that Kareena had liked to snort a little coke now and then, but if she was hooked up with heroin, I had no desire to get back into the picture with her. I'd just have to go on being lonely for a while.

"Looks like they were definitely a pair, man. Know what I mean?"

I shrugged. "What do I care?"

"Yeah. Well, guess what else? Go ahead, guess."

"I guess she won the lottery. For Pete's sake, Grady, just tell me, whatever it is."

"She had a baby bump, man, and it was real obvious. Maternity blouse, everything. I mean, she wasn't huge, but it was definitely visible, know what I'm saying?"

"OK, so she's going to be Mrs. Gus Maxton now, I guess. What do you want me to do about it?"

"Just thought you'd be interested to know, that's all. Hell, she won't marry the guy. Those guys never get married. They just father kids all over the place."

"So she'll have a little drug-addicted baby. And Gus Maxton will make his contribution to the human gene pool."

I was staring across the room at the vid case when I said this, thinking out loud. Grady's little crumb-snatcher was sucking noisily on Cassandra's plastic bottle while lying in a tiny pink and white blanket. The Luries' coats were hanging on the aluminum coat rack near the door, a slight dusting of snow still on them. I was always glad to see a little snow in winter. We hardly ever got any.

"Just thought you might be interested to hear is all," said Grady. "I'm not trying to hang your puppies or anything."

70

A Man Who Never Smiled

Nothing much happened for the next two months. Winter lasted for about a month, then it was back to iron gray skies, rain, and lightning storms. I watched Wild Beast World and worked at the brush factory pretty steadily. I ate real food whenever I could afford it, but I didn't want to blow through my legacy buying dinners. I went to a financial advisor to get some suggestions.

The financial advisor was a guy named Lightfoot, who said he had once been a schoolteacher. He was located on the second floor of a brown office building downtown. Inside, the building had green walls, but in the office itself, the walls were institutional cream. I sat in a cushioned chair and refrained from plugging my ears while Lightfoot talked because he had one of those booming voices like he thinks you're across a football field from him and you can't hear him or something.

I hate that.

"No use putting it in the stock market!" he roared. "There's no growth anymore! Too risky! Gotta be bonds!"

"What's the return on bonds?" I asked.

"Third of a percent! Was half a percent until a month ago!"

At those rates, I figured it might be fun to take my money out of the bank and stuff it in my freezer.

"No investment!" Lightfoot shouted. "Insurance rates too high! Insurers afraid of hurricanes and floods! Investors too insecure! Weak economy!"

I said thank you very much and got out of there before I lost my hearing.

It was Spring now, the season for refugees. They kept coming. Plenty of them didn't stay but moved on somewhere else, generally north. North was the direction of migration. Some of the courts had declared the 'unused bedroom' law unconstitutional and those cases were now being appealed.

I had learned to follow legal stories in the news ever since my troubles with the spooks had begun, and especially now that I was hooked up to the people in Akron.

Speaking of whom, they had me come in one day when I was able to get a cab ride up there. I was introduced to George Natale, the owner of Illustrious Ancestors Dot Com. Grady sat in the waiting room while I was ushered into the inner sanctum with Lumens and Phillips to meet Natale.

Natale was a pleasant-looking fellow, in his sixties, I'd say, with a charming lop-sided smile, brown eyes that blinked at you earnestly behind his glasses, a ski nose and curly, graying hair. He wore a brown blazer and black slacks, and a sport shirt with no tie. When he talked, he sounded like he was chewing on a walnut.

"Delighted to meet you," he said, shaking my hand. "I have the papers you signed, and I have your uncle's statement safely on file in the archives of Illustrious Ancestors."

This sounded so impressive, I half-expected someone to congratulate me.

"Are you very interested," asked Natale, his eyes so-earnest behind the lenses, "in genealogy?"

"I've never thought about it much," I confessed.

"Well, it's something that many people find an interesting avocation. Mormons, especially, are interested in it. You're not a Mormon, by any chance?"

"Nope."

"Well, you know, the Mormons believe that they will visit with their ancestors in the afterlife, and so, of course, they're interested in finding out all they can about them before they actually meet them."

Before any of us could react, a sharp, New York accent

launched its entry into the room.

"Founded by a schizophrenic fourteen-year-old from a family of rock-reading con men," it said. "Used his charismatic influence to construct a harem from his flock. Killed by a gunshot while jumping out a jail cell window."

We turned toward the door and saw in the doorway a very broad-shouldered man enter the room. He looked like a wrestler dressed in a shiny black business suit. He was medium height, with a balding, round head, a very thick nose and a small dark mustache. His fingers looked like baseball mitts. I guessed his age at between fifty-five and sixty. When he entered the room, Lumens and Phillips stood up, quite obviously out of deep respect. So then Natale and I stood up, also.

"True or false?" spat the wrestler in a sharp staccato.

Each of us looked at the other in bewilderment.

"True or false?" this time addressing Phillips.

"Are—are you talking about Joseph Smith, the Mormon prophet, sir?"

"The same! True or false, Phillips?"

"Sir, I have no idea. Suppose you tell us?"

"Can't do it, Phillips. I make no judgments. They are merely alleged facts, propositions. I deal in propositions, do I not?"

Phillips and Lumens nodded vigorously.

The wrestler continued. "Juries dispose. Judges dispose. I merely propose. I neither know the truth, nor concern myself with it. I am given a proposition. I argue it. Who have we got here?" he said suddenly, his manner shifting entirely to one of enquiry.

Juanita Lumens leaped in to perform the introductions.

"Mr. LaSevere, this is George Natale of Illustrious Ancestors," (shaking hands), "and this is Jeff Claymarker. This is Mr. LaSevere, the attorney who will be handling Tom Glendinning's trial."

"If we get one," LaSevere said.

"If we get one," echoed the law students.

I felt my hand swallowed up in LaSevere's warm baseball mitt and then released.

"These are my raw material," said LaSevere, his head twitching like a rooster's as he regarded first me, then George Natale. "You've met? You've spoken?"

"We've only just met," said Natale, "but I have his files in my archive."

"Good, good, good," said LaSevere, who, although apparently pleased, did not smile, and was—I could tell—a man who seldom if ever smiled.

"Do you," he said, turning to me an imposing eagle eye, "understand the importance of Mr. Natale to Tom's case, young man?"

"I—well, no, actually, I don't."

"Ahh-hh! Then my assistants have not informed you of the hearsay rule."

His assistants admitted that they hadn't.

"It is of crucial importance to your friend's case," said Mr. LaSevere, "that the jury understands there is a program used by American intelligence operatives to gather DNA from a water card in order to frame someone for a crime—yes?"

"Yes," I said.

"And how do you suppose the jury learns of such a program?"

"It's on my uncle's vid," I answered.

He cocked his large, thick-necked head to one side and eyed me as if I were a two-by-four being sized up for a bookcase.

"And how," he said, lowering his voice almost to a whisper and leaning toward me in an uncomfortable cross examination, "does your uncle's vid, as you call it, come to be played for the jury?"

"I—I guess you just play it for them?"

LaSevere lifted one of his mitts, and the two assistants chimed in together: "OBJECTION!"

"Sustained!" barked LaSevere with a pump of his fist and two strides across the room. "Sustained on hearsay grounds. The playing of the recording is not permitted."

He turned to me, eyebrows raised, and nodded his head once, as if to say, 'Do you see? Do you agree?' But of course, I had no idea what he was talking about. I said nothing, and didn't have to, because in a second he resumed his slow pacing back and forth, and his catechism of the two law students.

"And how do we avoid this lamentable result?"

"With the business record exception," Phillips piped in.

"What rule?"

"803(6)," chimed Lumens.

"Who is our authenticating witness?"

He held a finger in the air, aimed rather menacingly at his assistants, and they fell silent. Then the finger pointed at me.

"Who is our authenticating witness?"

I had no idea what he meant, but he obviously expected me to answer. I felt only a sinking feeling in the pit of my stomach, since I hadn't a clue what to say, until suddenly I saw LaSevere winking and nodding in the direction of George Natale.

I gave it a shot. "Mr. Natale?"

"CORRECT! We call Mr. Natale to the stand." Turning to the man in question. "Mr. Natale."

"Yes, sir?"

"I show you the item you received from Mr. Claymarker. It is—what, exactly?"

"A flash drive, sir."

"A flash drive. And it was given to you by Mr. Claymarker for what purpose, sir?"

"To be archived in my genealogy files."

"You are in the business of archiving such files at the request of such persons as Mr. Claymarker?"

"Why, yes, sir. And many other people besides."

"This is a business? You derive income from it?"

Natale squinted and worked on the walnut in his cheek. "A modest income, sir, but it is my business, yes."

"And the keeping of such records—the keeping of this record—is done in the ordinary course of activity at your place of business?"

"Yes, sir."

"Keep it on file there, do you?"

"Yes, yes, indeed I do."

LaSevere swung and addressed his assistants.

"I offer the exhibit in evidence!"

"ADMITTED!" they sang in chorus.

"Now," turning to me, "Mr. Claymarker, now I'm able to play the recording for the jury. And they will listen to every word. You understand?"

I nodded vigorously, even though I did not really understand at all.

"And now, my friends," said LaSevere in a much quieter, more conversational tone, "I must ask to speak to Mr. Claymarker alone."

Without a word, the two assistants and Mr. Natale left the room to join Grady in the waiting area and shut the door behind them.

71

Remember the Light

When we were alone, Mr. LaSevere sat in one of the chairs directly facing mine, placed his fingertips together, and spoke in confidential tones.

"You've struck the soft underbelly of the National Security state, Mr. Claymarker. You know that."

"Yes."

"You're aware that it's dangerous."

"Yeah, I'm aware of that."

"There have already been stories in the news about the claim of secret government programs for framing people for crimes."

I had seen them. "Yes."

"And so there's some light shining on us."

"Some light, sir?"

"Mr. Claymarker, it is not only we—or should I say, you—who must be careful, it is also they who must be careful. And they know it."

"What do you mean, sir?"

His tented hands fell, his chin sagged to his chest, he exhaled as he recalculated his approach. When his face rose, I could see he had chosen a path of frankness.

"Suppose, Mr. Claymarker, that you disappeared tomorrow."

To tell the truth, I had felt for some time now that I might well disappear at any time.

"There would be a story about it," LaSevere continued. "Wouldn't there?"

"I suppose so."

"It would seem suspicious, wouldn't it? Secret programs about to be exposed—and the key witnesses begin to disappear. You see?"

"You're saying they won't come after me now?"

"I don't say that. I don't go so far as to say that. I merely say there are things that others must consider. I'm saying that the danger can shift ground—here one minute, more distant the next. The game is public perception. There will be every effort to control public perception, if not by one method, then by another. But that is always the game, not simply who has the most players standing at the end of it. You see?"

"In other words, it isn't just a matter of silencing me because they don't like me?"

"It has nothing to do with liking or not liking you. Nothing to do with that! This is policy. These are weighty matters. We have Mr. Natale now. Mr. Natale is the custodian of your uncle's words. You are not the custodian. That's not to say that your testimony is of no consequence. It is. But the information is shared now. We have different angles leading into our propositions. That's the most important thing. And remember the light!"

72

Blow Holes and Seam Fires

People were dying by the thousands in Siberia. They were turning up asphyxiated. The cause was unclear, but in the news vids the talking heads were theorizing that it had something to do with the methane blow holes that had begun appearing early in this century, when the permafrost melted, and that had been releasing increasing quantities of methane into the atmosphere for decades. It wasn't the methane, they said, because methane wasn't toxic. They thought it might be a chemical reaction in the air that produced carbon monoxide, or else somehow decreased the oxygen content.

Evacuations were underway in the affected areas. Treatments for the sick were not even possible near the blow hole sites, because the effects were in the air and affected the care givers.

Pretty scary stuff, but not a whole lot different, in my mind, from how the old people down in the South would die off in the summer because they didn't have air conditioning. They would get heat stroke, things like that. It happened every summer.

At one point, I broke down and allowed myself to listen to a Wesley Wright broadcast, and he was pooh-poohing the whole Siberian thing.

"Siberia! Who lives in Siberia?" he said. "Have you read those Dostoevsky novels, those Tolstoy novels? People got sent to Siberia as a punishment, because nobody lived there! Oh, sure, people say, 'Wes, there's fifty million people living in Siberia.' Right. Do you have any idea how BIG Siberia is? Fifty million people is a drop in the ocean in a land mass of that size. Friends, don't worry about Siberia. Methane in Siberia will have the same fate as the Nazis, the same fate as Napoleon's army when it invaded Russia. It'll get swallowed up!"

The talking heads said that the methane releases were having effects way beyond Siberia, because methane had twenty times the impact of CO_2 as a greenhouse gas, although the CO_2 effects lasted longer.

It was a funny (strange) thing. I was viewing these stories on the net, and then I saw a comment from someone that said, "And yet the sun rises every day"—because, despite the continual cloud cover, the sky does get lighter in the day than in the night time—and there were dozens of 'thumbs up' numbers on the icon above the comment. I guess people like to be cheerful, and they don't like downers.

That's human nature. But it's kind of stupid, if you see what I mean.

In the meantime, in the state right next door to mine, Pennsylvania, the Earth had opened up near a former coal town, and a fire ran along the hole for five miles. The talking heads said the fire had been burning for about eighty years and would be hard to put out because it had to be fought entirely with helicopters. This was because they couldn't get near it from the ground—with trucks. It was burning on what they called a coal seam in the middle of the huge hole. They said the fire was so big that they planned to let it go on burning.

They were evacuating the nearby populated areas.

The presidential primaries were underway, and if you wanted to, you could spend hours every day watching campaign stops and speeches by the various candidates. I watched some of them to see if they had anything to say about these natural disasters. Mostly what they said was that they cared, and that they would do all they could to address the problems in an economically responsible way. None of them had any specific ideas on what to do.

I guess there really isn't anything people can do about these kinds of things. Is there?

73

Two Things Happen

The court of appeals ruled in Tom's favor, and now we were going to have a new trial. I was looking over my shoulder every day at the brush factory, and when my hours got cut back to half-time, I was real paranoid about it, but I kept on.

That's the way life is. You have to keep on until it ends.

Warner Phillips said Tom would soon be transferred to the local jail because he was no longer under a prison sentence, and that if I had a way to pay the bond, I could look into it and possibly spring him. But that would have to wait until Tom arrived back in town.

And before anything else happened on Tom's case, two big things happened. One was a certified letter I got in the mail from the Common Pleas Court Juvenile Division. When I saw it was from a court, naturally I thought it had to do with Tom's case and I didn't pay much attention to the "Juvenile Division," because I didn't know what it meant.

The letter inside told me what it meant.

It was papers from a judge that said I was ordered to take a DNA test to determine if I was the father of a child. It said to call a certain number, and when I did, I found out that the mother was Kareena Hecht.

I had to go down to a Children's Agency and give them a swab from my mouth. The agency was in a big brick building that looked like a haunted house. On the inside, it looked like the worst grade school you ever attended—green walls, locked doors in hallways, a reception area with a big window and a woman behind it who did not look up when you came up to it.

A lady took the mouth swabs with a Q-tip, and when I asked her questions, she told me to ask what she called a caseworker. I never found the caseworker, but I managed to find out that the baby in question wasn't even a week old, and that one test had already been done and the alleged father in that test had been ruled out.

I figured that must have been that Gus Maxton I'd heard about from Grady Lurie.

It wasn't long after this episode that I got another letter from the Court of Common Pleas Juvenile Division which announced that, as the result of the DNA test, I was alleged to be the father of a child referred to only as "S.H.," and that certain rights and responsibilities went along with this, and that there was to be a court hearing about it, and that I was expected to be there, and that if I couldn't afford an attorney, I could ask to have one appointed for me.

And the day I got that letter was the same day Irene Glendinning reappeared with little Tommy, on my doorstep, asking if I could give them a place to stay.

74

Alaska

I was blown away. I ushered Irene and Tommy inside. They had all their belongings in one half-sized flexible flight bag, which Irene set next to the door on entering. She was concerned about her wet umbrella, but I said, just leave it opened up on the floor and it would dry out. She looked pale and thin, but she refused food. Actually, I didn't have any real food to offer, but I gave Tommy some government issue wafers with bottled water.

Irene sank onto the sofa and sat there in a kind of traveler's shell shock. Naturally, I asked her all about where they'd been and what they'd been doing.

"We ended up in Alaska," she said. "It's beautiful up there, Jeff. You can see the sky more. But there are refugees all over the place. In Canada too."

"There can't be many more refugees there than there are here."

"Oh, yes, Jeff, there are two or three times as many. They have serious overloads, but it's still better up there, I think."

"Aren't there some methane craters there?"

She shook her head and shrugged. "I don't know. Maybe so. There's probably no place that's truly safe."

She said she'd spent some time in a homeless shelter in Anchorage. She had left because she couldn't stand the thought that Tom was a murderer and she only wanted to get far, far away. But, after she was up North, she followed the local stories on the universal vids, and she learned that the conviction had been reversed and Tom was getting a new trial. She didn't know about the State's attempts to stop that in its tracks, or how the attempts had failed, so I explained all that to her, to her great relief. Her main feeling was being upset with herself for having run off.

"Oh, Jeff, he'll never forgive me now. I lost faith in him. I didn't know what to believe! But I know now he didn't do it."

"He'll forgive you," I said. "There's nothing to forgive, Irene. I know Tom will take you back in a heartbeat. It's all he thinks of, all he wishes for, believe me. Just knowing you're back will give him so much hope."

"Do you think so?"

"I know so. Look."

I retrieved the letter he'd sent in which he said that Irene and Tommy were like the sun and rain to a flower, and that set Irene off. She was crying for at least ten minutes. Tommy got worried about her and started to cry himself before she regained control.

"Jeff, can you put us up until we can get on our feet? Just a few days would help."

I had a roll-out mattress pad and we put some blankets on that and Tommy slept on that in the den. Irene slept on the living room couch, insisted she wasn't going to displace me from my bedroom. I didn't tell her until the next day about my paternity issue, but that hearing was coming up in a few days. I told her we would try to phone the prison and leave a message for Tom in the morning, or else try to find out where he was.

75

Leak on the Net

It was really hard getting a message to Tom if you weren't his attorney, so I called Warner Phillips and asked him to do it. His reaction surprised me.

"I thought you'd be calling to cuss me out," he said.

"For what?"

"You mean you haven't seen? The 2025 subcommittee hearing recording has been leaked to the press. I just saw the first AP story on the web. The shit's going to hit the fan in about five minutes."

"What? I told you not to—"

"I had nothing to do with it, Jeff. That flash drive you gave me is locked in a safe here in our office. It hasn't gone anywhere. The leak had to be from inside the government."

"But that's—"

"Impossible? No, apparently not. Unlikely? Yes. But it's happened. I swear to you it did not come from me. I would not be doing that right before our trial, trust me."

"What's going to happen now?"

"I don't know. I mentioned it to Mr. LaSevere, but he's focused on the trial. Jeff, in a weird kind of way, maybe it's a good thing. The subcommittee's a much bigger story than Tom's murder case, from National Security's point of view. Maybe it'll take some of the heat off us."

I was dumbfounded. There was no point in discussing it until we could see what was playing out in the media, so I returned to my original reason for calling.

"Can you get the word to Tom that his wife and kid are back and staying at my house, Warner?"

"Jeff, Tom is in your local jail now. You can go see him

yourself on visiting hours. I can give you the name of a bail bondsman, if you think that'd help."

I said that would help a lot. He gave me the name and number and I cut short the call. I was on my break at the brush factory and didn't have much time, but I was itching to get Irene hooked up with Tom, and possibly get him out of jail. And I was itching just as much to get some more time to myself, so I could follow the leak on the net.

76

Juvenile Court

And right in the middle of this three-ring circus came my paternity hearing in Juvenile Court.

I was becoming familiar with the layout at the courthouse, but I hadn't been in the Juvenile courtroom before. The rooms were up on the second floor. I found myself sitting at a big square table in the center of a courtroom, with three walls of chairs behind me, two ladies sitting across from me, a tall, bushy-haired man in a brown suit sitting in one of the chairs against the wall, and a lady judge sitting up high above us in a black robe. On the walls were oil portraits of past Juvenile Court judges, and they peered down at us as if imagining what they would do if they were here conducting the proceedings.

Kareena wasn't there.

The judge knew who we all were, and she announced it at the beginning of the hearing. One of the ladies across from me was the prosecutor, and the other one was a caseworker from Children's Services. Her name I remember: Renata Crossman. I would get to know her later. The tall, bushy-haired man in the brown suit was somebody from the Child Support Agency. I was what the judge called the 'alleged father'.

"And where is the mother?" the judge asked the two ladies on the other side of my table.

"Still in the hospital," answered the prosecutor, a rather small, auburn-haired woman with round eyeglass lenses.

"In the hospital because—?"

"Being treated for heroin overdose, I believe," said Renata Crossman. She was a thin, coffee-colored lady. "She's stable, so far as I know, but she couldn't come today."

"Well, we're here primarily because of the alleged father,"

said the judge, whose name was on a credenza in front of her elevated platform. "Judge Camp," it said. "Mr. Claymarker, before we go any further, I must inform you that you have the right to have an attorney represent you in this matter. What we are here for, just so you know, is to determine whether or not you are the father of the child born to Ms. Hecht."

"The DNA test made that pretty clear, I thought," I said.

"Well, before you say anymore, I need to tell you that if you're unable to hire an attorney, I can appoint one for you at no cost to yourself, provided you're income-eligible."

"Just to decide if I'm the father, you mean?"

"Yes. And for any matters dealing with your relationship with the child if you are determined to be the father."

"Like what? You mean custody or something?"

"Yes, or visitation, or child support. But we can postpone today's hearing, if you want to engage an attorney."

"I don't need an attorney just to determine if I'm the father. There's a DNA test."

"Are you saying that you wish to proceed today without an attorney? Because if that's what you want to do, you can do that. But you must understand that a finding that you're a parent entails responsibilities as well as rights, including, as I mentioned, the duty to pay child support, in some instances. But we're only here today to answer the one question, whether or not you're the father."

"If that's the only question for today, I don't need an attorney for that."

"But, you're saying, beyond that, you would want an attorney?"

"Yes."

"But you're OK with going ahead today without one?"

"Yes."

"And do you dispute that you're the father of the child?"

"No. I've got a DNA test that says I'm the father and it's 99.9

percent certain."

"So you admit that you're the father?"

"Yes. And I'd like to know my child's name, if I may?"

"It's Shannon Hecht," said the thin caseworker.

"Do you mind if I ask you," said the judge, "about your relationship with Ms. Hecht. Is it Kareena?"

"Kareena," said the caseworker.

"Have you been in a relationship with Kareena?" asked the judge.

"Yes, I have," I said, although it seemed obvious to me that I had been.

"Has that been ongoing, or—?"

"We broke up end of summer."

"What date, or what month, if you recall?"

"End of August, or early September, around there, I believe."

"And before that—?"

I went over the history of the relationship as well as I could remember.

"Well, the dates work out," said Judge Camp. "And you willingly acknowledge that you are the father?"

"Yes."

"And what would you like to have happen with regard to your daughter?"

"I'd like to see her. I'd like to take her home, if Kareena's in the hospital. Where is she now?"

"She's in foster care," said Renata Crossman.

"She was removed for her own safety," the judge explained. "There was heroin in her system at birth. Visitation for you will be arranged by the agency and made a court order. As far as taking her home, that's a custody issue, and you may want to have an attorney to help you with that. Ms. Crossman," she said, addressing the caseworker, "are there any criminal actions pending?"

"Yes, the boyfriend's already been charged with trafficking, and I believe there are some kind of charges coming down for Kareena pretty quick here."

"Kareena's being charged with a crime?" I asked.

"Possession of heroin is a crime." The prosecutor nodded.

"I'm going to make a finding that Mr. Claymarker is the father, based on the DNA test and the acknowledgement. Mr. Burnside, you can meet with Mr. Claymarker to determine his income and submit a recommendation to the court."

The tall, bushy-haired man nodded.

I hadn't thought we were going to get into the question of my income just yet, but I didn't say anything about it.

"Mr. Claymarker, if you'll request an Affidavit from the Clerk, and fill it out, I'll appoint counsel to represent you as necessary. Will there be anything else?"

No one thought there would be anything else.

77

Dutch Is What They Speak in Holland

Tom's bond was a hundred thousand dollars cash or surety. Irene and I went to see the bondsman, whose name was Mortimer Sprigg. His office was on the east side of town, in a strip mall near an ALDI's. He was a small, pot-bellied man of about fifty with bug-eyes and a florid face, who looked like nothing so much as a contented frog sitting on a lily pad. He wore a beige sweater-vest with a navy blue suit and sat behind a desk holding his fingertips pressed together.

"Co-llateral," he said. "Co-llateral is what you need. Have you got co-llateral?"

"I've got nearly fifty thousand dollars in the bank," I said.

"Fifty thousand? That's good, but dollars move. We don't want things to move."

"My aunt has a house," said Irene.

"Does your aunt love you?"

"I think so."

"Because the aunts, they helps the ones they love."

"I think she might help," Irene said, eyeing Sprigg strangely.

"Well, a house now. A house don't move so easy. You thinks your aunt will help, then?"

"Yes, I hope so."

"Hope is not a signature. We need more than hope. Has you talked to your aunts, Miss?"

"Not at length, but she knows I need help. I think she'll help me."

"Then I thinks we can do business."

"What kind of accent is that?" I whispered in Irene's ear.

"Dutch," she said out loud, then blushed with embarrassment.

Sprigg appeared perplexed. "Dutch? What is Dutch?"

"We're in Dutch," I said. "How soon can we get Tom out of jail?"

"Well, first things firstest. Firstest, you convey a lien on the bank account, and secondest, you gets me together with the aunt."

"How soon can you get with your aunt?" I asked Irene.

She thought it would take a day or two.

"Once we gets those together, I goes and gets Tom out," said Sprigg.

We left the office and returned to the parking lot where a cab awaited us. The buildings all had a kind of pink paste color, except where they were lathered in logos. We had some unusual sunshine today, and it seemed like a good omen.

"What's wrong with the way that man talks?" Irene wondered.

"You said he was Dutch. Did you know what you were talking about?"

"No. I don't know why I said that."

"Have you ever been to the Netherlands, Irene?"

"Where's that? Dutch is what they speak in Holland."

78

One Big Armed Camp

The troop levels were still ultra-high down in Texas when the crowds began to swell in front of the Capitol in D.C. There was a lot of anger stirred up by the stories in the vids of how Congress had had before it the information about the weather patterns back in 2025, and had known it could do something then, but had declined to. In other words, the press leak of my uncle's secret recording of the Senate hearing had stirred up the hornet's nest of protesters. And I thought, sooner or later they're going to have to call in the troops up here, too. Pretty soon, America will be one big, armed camp.

That's what I was thinking, and it turned out to be right, because the crowds kept growing every day.

But before they reached their peak, that peculiar verbalist, Mortimer Sprigg, got Tom released from jail.

Irene went with him to get it done. She noticed, when they paid the court clerk, that Sprigg spoke to the clerk in normal language—none of his 'Dutch' talk. Then he turned to her and said, "You're wondering why I'm not speaking strangely."

"Well, you did before," said Irene.

"You remembered me because of it?"

"I guess I did."

"The ones that is successful is the ones that gets remembered."

And then he laughed like Popeye the sailor, Irene said. "Uk-uk-uk-uk-uk!"

Tom went home with Irene and Tommy, who by now were staying in a room at the St. Adolphus Hotel, but afterwards he dropped in on me, because we had agreed he should use my address on his bond, since it wasn't likely to change abruptly. He stood in my doorway, a strange figure in that space. Then he gave me a hug with tears in his eyes, and I went and got him a beer, because I had a couple of those on hand. We figured it didn't much matter if he stayed at the St. Adolphus, because we had wrist phones and could keep in touch 24/7.

He looked sallow and haunted, but at the same time ecstatically relieved.

"I've got them back," he kept repeating. "I've got Irene and Tommy back. I would have died without them, Jeff. I swear to God, I would have died."

"It's all right now, Tom."

"Is it? You're sure I'm not going back to hell? You'd bet on that?"

"I think LaSevere looks like a good lawyer, Tom."

"Yeah, I met him. He came to see me in prison. But what's this I hear about you being a father?"

I told him about Shannon, and Kareena being in the hospital, and the hearing I'd had.

"I get to visit her today," I said.

"Kareena?"

"No, Shannon. My kid."

"Wow. First time?"

"Yep."

"It'll blow your mind, Jeff. The first time you see your own kid, it changes your life."

"That's what I hear."

"You'll fall in love."

"Oh, I don't know. I'll probably look at it and say, 'Well, that's a baby.'"

"You'll fall in love, birdbrain. Trust me."

He turned out to be right.

79

Japanese Snow Forests

Shannon looked like a little monkey—one of those hooded, buff-colored monkeys in the Japanese snow forests. She had a bit of wispy brown hair floating about her head in a way that reminded me of Kareena, and she had features a little like Kareena too—but she had my eyes, and that combination made her look different from either one of us.

Renata Crossman handed her to me at the Children's Services building—that depressing old brick affair in mid-town that looked like a miniature version of every run-down elementary school you've ever been in, right down to the puce walls, the child-drawn murals of lakes and round trees, and the Eleanor Estes paperbacks lying on display tables. Shannon was wrapped in a gray blanket. Renata handed me a bottle and said, "It's feeding time."

I sat there on a folding chair, in a room without a picture on the wall, and held my daughter as she quietly nursed on her bottle just as natural as anything.

And I did, I fell in love right then and there.

"I've got to have this baby," I said. "I've got to have my baby."

Renata Crossman laughed, her thin frame rocking.

"Seem like she likes you," she said.

I didn't want to leave her, but I was only allowed two hours. Thankfully, Shannon was asleep when the foster mother took her away. I felt like crying.

"What do I have to do?" I said to Renata. "I'll do anything."

"We haven't even had an adjudication yet," she said.

"What's that?"

"That's when the judge declares the baby dependent. And then we do a dispo, and we give you a case plan. You got to work your case plan. And Mom's got to work hers, too."

"You mean Kareena?"

"Mm-hm."

"Are you going to give my baby back to Kareena?"

Renata pursed her lips and raised her eyebrows.

"Oohh, it's too soon to be saying who the baby ends up with. But looks like Kareena's got some problems, with the drug use and all."

I don't understand drug use. I'm too bourgeois. All I wanted was my daughter, and I could have married Kareena right then, if I'd thought it would mean I could keep Shannon. But Kareena had that penchant for getting high. Even when we were together, she was always wanting to snort cocaine. I did it a couple of times out of curiosity, but it didn't do anything for me. It was like Kool-Aid hitting the inside of my nose, and I didn't feel any particular effects from it. But Kareena was always wanting to snort. And now she was in the hospital, or—

"Is she still in the hospital?" I asked Renata.

"I believe she's out now."

As I was walking away, Renata called me back.

"You know I'll be wanting to check out your home, Jeff. You got baby supplies in it?"

"What? You mean like diapers?"

"Diapers, formula, bottles, play pen, toys, clothing."

"I'll get them," I said. "I'll get all those things today."

She laughed, but I was already on my way to the nearest big box store, having no idea where else to go.

80

I Would Have Pushed the No Button

If somebody had said to me, "You can have a child or not: push this button for 'yes', push this one for 'no'," I would have pushed the 'no' button. The condition of the world was not so great, not that it had ever been that great in my memory. But things kept getting worse and worse. The protesters at the Capitol in D.C. were massing in greater numbers every day, and the government had begun cordoning off the buildings and stationing troops in the area. Down in Texas, it was a fortress state. We were lucky if we had twenty clear weather days in a year. The sea levels kept rising. The strange deaths around the world kept growing in number. The food we had to eat was minimalist, to put it mildly. The economy continued dragging along at near-Depression levels. The refugees kept overflowing in the northern cities; the southwestern cities were dying off, burning up, drying out. People were in a panic. The only really good news was that the CO_2 levels in the atmosphere were flattening out, because we were finally running out of oil and gas. The rickshaw was making an appearance in populated areas, and there were more and more horse buggies, not only Amish ones. That was almost pleasant, yet it was a sign of desperation.

Some world to bring a kid into.

But since no one had asked me to push a button, and since my daughter was already here, I felt just like any normal father would. I wanted to take her to my heart, and raise her up, and help her grow to be a fine person, and watch her succeed in life better than I had.

We finally had our days in court and I started working my case plan with the agency. I saw Kareena in court, and she looked worse than I'd ever seen her. She stared into space, seemed depressed, would hardly talk. Her boyfriend, the infamous Maxton, was in jail, unable to make bond, charged with trafficking heroin in large quantities. Renata Crossman didn't seem to think he'd get out any time soon.

And Kareena was charged too, but they said her charges weren't as serious and she'd probably get probation. She said she wanted to work her case plan too, but it was easier for me, because I didn't have all the drug counseling and testing requirements she did. We were only a few weeks into the regime when Renata arranged for Shannon to visit in my home. By then I had all the necessary items — playpen, diapers, formula, etc. I was only working twenty hours at the brush factory.

Irene had regained some of her previous feistiness with regard to me, but it was more playful these days and less hyper-critical. She was happy to help provide moral support for my newfound parenthood, as was Tom. And she had plenty of advice on child care. So that helped a lot. When I wasn't working or taking care of Shannon, I was attending parenting classes, primarily to make points with Renata and her crew at the agency.

The days rolled on into winter, but there wasn't much snow. The leaves fell from the trees and lay in black puddles in the streets. The limbs rose up into the air and seemed to say, "I'm naked — naked and helpless."

Tom, meanwhile, had a meeting scheduled at the court. They called it a pretrial. I was at work when he went to that, but when I got home, he and Irene and Tommy were sitting on an old beat-up sofa that I kept on my covered front porch.

Tom and Irene were laughing.

81

He Flipped

I was coming up the walk when Tom and Irene spotted me.

"You guys are sure in a good mood," I called out. "What happened at court?"

Tom exploded with joy and ran down the steps, clasping me in a powerful bear hug.

"It's over, Jeff! I'm free! They dropped the case! I'm free! It's over!"

I was totally floored, couldn't believe it.

"What! What are you saying? It's over? No trial?"

"No trial, Jeff! The prosecutor asked to dismiss the case and the judge dismissed it! He said, 'You're free to go.' LaSevere shook my hand and said 'Bye-bye, baby!'"

"But how? What happened? Why did they drop the case?"

"That little bastard that held us up confessed to the murder! LaSevere said, 'He flipped.' Confessed to the whole thing!"

"Are you serious?"

"Confessed, Jeff! They're prosecuting him, not me. I'm exonerated."

"But what about the DNA?"

"I don't know and I don't care, man! We came straight over here to wait for you so we can go out and celebrate. I've got a little bit saved up. We can go to a restaurant. We can go to a fruit speakeasy. You pick a place, brother, and we're going."

"All right. Well, I'm springing, though. This is your night, Tom."

We did the town too. We went to a local restaurant and ate real food and drank a bottle of wine. We couldn't stay too late, because Irene had to put Tommy to bed, but it was a memorable evening—and such an incredible relief, you can't imagine, even for me, but much more so for Tom and Irene.

I had quite a few questions, though, and the next day I called Akron to see if I could get some answers. Neither Warner Phillips nor Juanita Lumens was in, but I was put through straight to the Big Man—Mr. LaSevere.

82

The Sky Is Orange Crush

I couldn't see LaSevere as we talked. It wasn't a skype call, just speaker phones. I imagined him in wrestler's tights, with the leather headgear, because fantasy was the new thing. Everywhere you went people were saying things like, "The ocean is lemonade. The sky is Orange Crush." I think the human race was going batty.

"Mr. Claymarker! How are you, sir?" came the tight, discerning East Coast-accented voice that didn't quite go with the wrestler's body.

"I'm fine, Mr. LaSevere. Tom told me that his case had been thrown out."

"Dismissed. With prejudice. Which means it can't come back. He's off the hook, and you will not need to testify, nor will Mr. Natale."

"So I gathered."

"And I'm told you had a question or two?"

"Yes, sir."

"You want me to explain the process to you. Explain to you exactly what happened. Yes?"

"If it's all right, sir."

"Perfectly happy to. We came to court for pretrial. The prosecutor announced that Mr. Milteer had made a complete and full confession to the crime. The prosecutor therefore asked the court to dismiss the case. Result: case dismissed."

"But—did they say why they were doing that? What about the DNA evidence?"

"The prosecutor has complete discretion, Mr. Claymarker. The proposition has now changed. He has a confession. He has the responsible party. He recognizes a mistake was made. He does not seek injustice. He asks that Tom be discharged. Ergo, Tom is discharged."

"Well—did that kid—Milteer—admit that the government had set Tom up?"

"That was not discussed. Please understand, Mr. Claymarker, that I have not read a description or a transcript of the gentleman's confession. My understanding was that he claimed responsibility for the whole thing."

"But Tom was framed."

"And I assume Milteer was responsible for that."

"But that was a sophisticated method that was used. It was known only to people in National Security, in Washington. It was top secret."

"Mr. Claymarker. Do I detect a certain disappointment in your concerns? Do I detect a certain preoccupation with a little thing called the truth?"

"I'm wondering about the truth, sir, yes."

"You are not my client. You need not take advice from me. But if you were my client, I would advise you to focus not on truth, but on propositions. The proposition before us, prior to yesterday, was that Tom had brutally murdered a young woman. Today, the proposition is that another young man did it, and Tom had nothing to do with it. That is a much better proposition, is it not?"

"But it lets the government off the hook."

"Would you rather go to trial with your friend's life on the line?"

"No, but—"

"Listen to me, Mr. Claymarker. Don't imagine that I'm naive about what you're saying. And don't imagine that the story you're concerned with is forever unavailable to historians or other interested parties. You'll recall that a few months ago, the story in the news sources was about how your friend had asked for a new trial on the grounds of new evidence showing the existence of certain top secret government programs used to set up unwitting fall guys for criminal acts. Then the new trial was granted. That's out there. That's available to someone who researches it. Now, of course, the story has changed. Now the story is about a young murderer who went undetected and finally came forward and confessed. And there will be stories about his motivation, or his lack of it. I daresay there will be stories about his insanity. And I wouldn't be surprised if he puts up a defense of insanity. Because a confession is one thing, and an admission of legal guilt is quite another."

"How can he claim insanity if they believe his confession?"

"Oh, he can do that. He can say, 'I was crazy then—not now.' But, in any event, you're right that the story has changed, and changed considerably. And don't think, Mr. Claymarker, that I haven't considered all the ramifications of that. Don't think, for example, that I haven't considered that someone may have wanted Milteer to confess—and to do it precisely to change the nature of the story that gets bandied about in the press."

"But you're talking about people who are mixed up in it. Are those people going to be held responsible?"

"I very much doubt it, young man. Who is interested in them? You? Me? The State of Ohio? Well, I guess you are, more than anyone else. Because you're interested in that quicksilver essence, the truth. But the truth is not what appears in the news sources. It's hidden away. It flashes near the surface once in a while, like a fish rising to suck up a water fly. But then it descends again, down into the deep. And if someone wants it, they have to look at yesterday's stories, the stories about the reasons for the new trial, and they have to put two and two together. They have to swim down into the deep waters. And so few people take the trouble to do that. They want their news pre-digested, served up pat, tied up with ribbons and bows. Forgive me for mixing metaphors, but you take my point. The truth is not a thing for Everyman, Mr. Claymarker. Everyman doesn't want to do any work to get it. And as for myself, if I concerned myself with the truth, I'd never have the time to move on to the next case. I'd be mired down and soon sunk. So I stay focused on propositions—the favorable and the unfavorable, the moot and the active, the decided and the undecided. It would not be an unhealthy thing if you did the same. Remember that a short while back you and I were seriously concerned about your safety. As things stand, I think your safety is in a much better position going forward—at least if you can put the truth out of your mind."

After talking with LaSevere, I had to admit that some things were much clearer—like the fact that I was the only one who cared about the truth coming out. That was important to realize, because I knew better than anyone that I was nobody. And that meant that Nobody was interested in the truth.

83

Quicksilver and Sucked Air

The public furor over the lies and actions taken by the government going back to 2025 kept growing. The crowds near the Capitol were numbering over a hundred thousand, and they were protesting 24/7. There were military troops escorting the Congress people in and out of the building, and threatening the protesters to stay behind the cordons on pain of being tear-gassed. Still, the crowds continued to grow.

And as they grew, the media was plastered with a story about a woman accused of the leak of my uncle's surreptitious vid. She was a government agent named Adrienne Raines. The first photos of her on the net showed a young, blonde woman, wearing large shades over her eyes, hair in a chignon. She was currently residing in Switzerland, and the U.S. government sought to have her extradited and tried in the U.S. for treason.

It must have been the third or fourth time I saw a photo of her that I realized I'd seen that mouth before, that nose, that face. And then I realized I was looking at Nola Sheeka. She had completely altered her appearance, using all those techniques women have at their disposal for that purpose. But who had been in a better position to leak my uncle's vid than she? No one.

As to why she had done it, I could only speculate at that time, but I could not imagine another reason aside from that lunatic desire for truth that Mr. LaSevere had considered of so little importance. It seemed that, underneath her devotion to duty and national security, Nola—or Adrienne—had an impulse toward that very quicksilver essence, the truth. The impulse had gotten the better of her, and now she was a hunted figure.

The political season was underway. We'd had the Iowa caucuses and the New Hampshire primary, and the focus was on the Republicans because they were the out-of-power party in the White House, but there was no clear front-runner so far. The big story was really about the strong showing of a remarkably uncharismatic candidate running from a third party, the National Security Party. His name was Cornwell, but his name doesn't even matter, because no sooner had he pulled in thirty percent of the vote in New Hampshire than my old friend Wesley Wright declared himself a candidate for president from the same party.

And Wesley Wright soon sucked all the air out of that room.

84

Renaissance

"My friends," said Wesley Wright in his standard stump speech going into the Michigan primary, "our nation is in crisis. Americans are under siege at home and abroad. In the nearly four years he's been in office, President Goya could not put away the Middle Eastern Caliphate, in spite of having the most powerful weaponry and the biggest and best-trained military machine on the face of the Earth. And make no mistake, my friends, despite all the claims that we've become a second-rate power, there is still no military power on Earth comparable to the United States of America. This has been true now for a hundred years, at least. And what we have had in the Middle East now for half a century is a drain. It's a big sinkhole and it's sucking up all our blood and treasure. And it's time to stop. And we can stop it tomorrow. All it takes is political will. And what you see in these gatherings wherever I go, what you see in the rise of the National Security Party, is a gathering of political will. It's the will of patriots all across our land, Americans who say, 'Enough of the Republicans, enough of the Democrats, they are nothing but Tweedledum and Tweedledee.' We are at war, my friends. War. And there are no half-measures in war. In war, you fight to win. You don't hold back. Well, we have been holding back now for a century, and a century is long enough. We need to go in and nuke the Caliphate, erase them from the map. Then we will start over with people of good will.

"And as for the state of siege here at home. You leftists, you Marxists, you haters of free markets and free ideas, I know what you have to say. You say that people are outraged at the selling of our country. You say that Big Oil and Gas bought our lawmakers twenty-five years ago, or fifty years before that, and they covered up what they knew, and they could have saved us from the world we live in today, with refugees coming from the desert of the West. You say they could have saved us from the black and orange skies and the endless rain in the East, and the horrible, terrible carbon dioxide emissions that you think have led to all the weather-related anomalies we've suffered now for fifty years. You Greens think capitalism caused the weather! But I've got news for you. God causes the weather!"

(Cheers interrupt the speech at this point.)

"You think New York is on stilts because of global warming, and you think Man caused it. You don't have any perception of natural cycles. Did you know that the Earth had no polar ice caps back in ancient paleolithic times? Millions of years ago, no polar ice caps! And then there was an Ice Age. You pathetic little Swedish style socialist advocates don't realize that Nature is bigger than Man is! And you want to blame everything on Capitalism. What about Consumerism? Did you stop driving your gasoline-powered car before we ran out of gasoline? Did you buy those little wind vehicles that stop running after thirty miles back when it might have made a difference? No. You are hypocrites!

"So you ask me, what will I do about the protests in Washington? My friends, I will do what needs to be done. I will bring a return to Law and Order. I will not have scruffy malcontents defacing the hallowed halls of Congress with their picket signs and their loud mouths. I will clean that situation up just like you would clean up garbage dumped on your front lawn.

"But I'm not going to spend my time before you being negative. You want to know what I stand for. You want to know what I'll do when I become leader of the free world. I'll tell you.

"I'm going to refuse to sign any budget from Congress that does not cut back social spending by fifty percent. And I'm going to refuse to sign any budget that does not increase defense spending by at least that much. And I'll tell you why.

"As your Commander-In-Chief, I plan to save our nation untold billions or trillions of dollars by eliminating the huge yearly expenditures for the prosecution of war against the Caliphate and our enemies in the Middle East and elsewhere. I will do this by ending the war. But I won't end it by pulling out. I'll end it by destroying our enemy. I'll end it by winning the war."

(Cheers overtake the speech at this point.)

"We need a strong Commander-In-Chief. We need a strong leader at home and abroad. We need to consolidate power and place it where a strong leader can use it, which means in defense, because the president is not Commander-In-Chief of any other department or sub-department or sub-sub-department of this massive, Hydra-headed social welfare government which has plagued us for over a century. And when I'm your leader, I'm going to dismantle this bloated monster of a government stick by stick, and return all the power to you, the people. Schools will be governed by local communities, not by an Education Department in Washington that knows nothing about your community or your child or you. Health care will be controlled by doctors and patients, not by bureaucrats with a one-size-fits-all mentality. No EPA in a Wesley Wright administration will be telling businesses and energy producers where they can locate, what they can emit or not emit, how much they can build, how many jobs they must cut to save an endangered rodent.

"The American businessman will be set free in my adminis-tration. Free to invest, free to grow, free to employ people, free to make our economy vibrant again, instead of sluggish and fearful and intimidated by government bureaucrats and I.R.S. agents.

"If you agree with me, friends, then give me your vote, and America will be on the way to a Renaissance, and we will lead the world to a bright and vibrant new day!"

The people listening to this speech looked like members of a motorcycle club. They'd be wearing sleeveless, cut-off shirts, and there'd be gray-haired men with ponytails, women wearing bowling shirts, and fat mothers pushing strollers. Most of them were angry-looking white people of all ages. In short, they looked like a lot of Americans.

Wright got more votes in the Michigan primary than anyone else, beating the second-place guy by ten percent. The second-place guy was the other National Security Party candidate.

85

Exposé

Tom could see where all this was heading—or he said he could.

"We're becoming a fascist dystopia," he said.

Irene and Tommy played on the sofa with Shannon, who was now staying with me for four hours at a stretch without agency supervision. I played too, dangling a rattle over her cooing face and outstretched tiny arms, while listening to Tom rant as he paced nervously up and down the braided throw rug on my living room floor.

"Either that or everything will simply break completely down and a twenty dollar bill won't be worth anything. And the air will keep getting worse. Jeff, we're thinking of heading back North."

"Back North?"

"Alaska. Where Irene went when I was in prison. She knows where to go. Buddy, you're not working that much. Why don't you come with us?"

I knew very well why I couldn't. "I've got to finish doing my case plan and get custody of Shannon."

"OK, but when you're all done with that. What about we take off and go North then? You, me, Irene, Tommy, Shannon?"

"Kareena might not like it. It's her daughter, too."

"Hell, she can come with us."

"She won't want to go with me."

"She doesn't have to be with you. She can come as a friend among friends. Better deal for everybody."

Irene spoke up, reverting to her old skepticism toward me.

"There's no guarantee he'll get custody," she said, addressing herself to Tom. "Kareena might get it. That's mother-daughter."

"Oh, she's too fucked up," said Tom. "The woman's been

mainlining heroin for at least six months. The poor kid had it in her system when she was born."

Shannon had, in fact, been born addicted, the doctors said. Her symptoms had been remarkably light, and she had been recovering well, so we were lucky. But when I thought about her being born in that condition I became angry and resentful toward Kareena. I guessed the 'old' Irene would have said that Kareena's addictions were partly my fault. Hadn't I 'used' Kareena? Hadn't I been unwilling to commit? I knew I was no saint. But Kareena had been a druggie ever since I'd first known her, and neither one of us had ever contemplated having a child. As far as I knew, Kareena had always used an IUD. I was no better nor worse than plenty of guys my age, but having Shannon changed the landscape. It changed it so much that I thought seriously about going and having a talk with Kareena, about maybe accompanying Tom and Irene up North.

I didn't have that talk with her right away, though. I was afraid if I did, that she'd waltz into court and announce I was planning to leave the state or the country, and then the agency might not want me to have custody.

Being stuck in the courts all the time left you not trusting anybody.

Because of that lump sum I had in the bank from my uncle, I wasn't allowed a court-appointed attorney in the Juvenile Court case, so I hired one. Her name was Courtney Breit. We went to court a couple of times, and when I asked her for advice she said, "Just sit tight. You're doing what you need to do."

I asked her, "Should we ask for custody? Can we do that?"

"We can do it, but the court won't grant it unless the agency asks for it. Just keep doing everything right and the agency will ask for it."

That seemed really easy, and I figured I could do it myself, so I fired Courtney Breit and represented myself. I had to save some funds somehow.

There was a lady appointed by the court, called a Guardian ad Litem, who was supposed to keep tabs on Shannon and write reports for the judge about how things were going, and I got on her good side right away. Her name was Prudence Miller. She was young, about my age, and nice-looking—dark hair, green eyes, deeply tanned skin, an interesting look—but newly married, so don't get me wrong. When I say I got on her good side, I mean that I kept in touch with her and saw that everything went well during visits with Shannon that she sat in on, kept my home clean and neat, and went to my parenting classes. Pru seemed pleased with that.

Meanwhile, Kareena was convicted of heroin possession and ordered into a rehab program. She had to live in a home with other addicts and attend drug counseling classes, things like that. She only saw Shannon two hours per month. They had to bring the baby to the halfway house for that.

Everyone could see the writing on the wall, even Kareena. Kareena told Prudence Miller that I should get custody. It took another three months for that to happen, but suddenly I had legal custody of Shannon and the case was closed, shut down. Kareena was to have 'reasonable' visitation, which meant there had to be some, but she and I could agree on what it was. At the time I was granted custody, Kareena was still in rehab and expected to be there for another couple of months.

Meanwhile, the crowds in D.C. had kept getting bigger and bigger. The military people unleashed tear gas attacks almost on a weekly basis, but after the dispersals and the clean-up, the crowds came back bigger than before. Finally, they stopped tear-gassing them. They had permanent cordons set up, theoretically so that traffic could move through the streets, but the plan wasn't working so well. The crowds chanted, sang—and grew. They carried signs reading YOU KILLED MOTHER EARTH or AIR & WATER: SOLD FOR OIL & GAS or 30 PIECES OF SILVER or YOU'RE RICH, I'M HOT. The talking heads on the vids were saying that the protesters made up the twenty-ninth largest city in the country—hundreds of thousands of people. Some Congress people wanted to pass legislation to clear them out, but they couldn't get a majority of votes for it. Congress was actually afraid of the people for once.

And the election speeches went on. The self-styled patriots among the candidates were fond of attacking Nola—I should say, Adrienne Raines. She was now in Russia, seeking asylum, and the Russians weren't turning her over. At least that cat was out of the bag and I was no longer the culprit for it. As for that other bagged cat, the spooks had that one under control. I mean, the story about the murder of Teresa Pagano, aka Hypnotica Christiansen.

I saw a story about it on a program called Exposé. The talking head on Exposé said that Tom had been wrongly accused, but it said he had gotten back into court on a 'wild claim' that he had been framed by 'unknown secret government agencies', whereas the real killer finally came forward and confessed, 'clearing the whole thing up'.

Some exposé, huh?

I noticed that no matter which channel you went to on the networks, this was the only version of the story you could get. Then I went back and did some reading and learned that ever since the 1950s, the CIA and the agents of national security had controlled the major media. They had a program way back then called "Operation Mockingbird," where they put what they called human 'assets' into the media, or developed relationships with reporters so that there was a reliance on the agencies for information, and in return the reporters gave the news the proper slant.

I'd heard about this kind of thing before, but now I was seeing it, blatant and omnipresent, all the airwaves and the print media filled with lies.

And, just as LaSevere had predicted, Guy Milteer pled Not Guilty By Reason of Insanity.

86

Hurricane Bruce

And then the people broke through. They forced their way into the halls of Congress, into the Capitol building, and took over all the desks. A couple of them were shot by armed guards, but the guards were overpowered. The injured were taken to hospitals. None of them died. The military was strangely quiet. It reminded me of the old vids about Tiananmen Square, the man standing in front of the tank, moving about to stay in front of the long barrel, except that, unlike Tiananmen, there was no major massacre. It was one of those moments in history when everyone seems to know that a change has to occur, will occur, must occur. The news coverage of the event was 24/7, masses of people inside every room in the Capitol and massed outside on the steps, in the street, in the park; military men looking on, the President giving speeches, calling for calm, trying to put a positive spin on it ("The people want their voices to be heard, and they should be heard. I hear those voices.")

But we all knew the lights would be going out soon. They'd go out one by one, or they'd go out all at once, but they'd go out.

And Tom came back with his mantra.

"It's time to go, Jeff. You've got custody now. America is shutting down. Closed for business. There'll be more refugees now than ever. Money won't be worth anything. It's time, man, can't you feel it?"

I could feel it.

So could Grady Lurie and Cassandra. They now wanted to join Tom and Irene in the trek northward. Everyone was waiting for me.

I made an appointment to visit Kareena in her dry-out tank. They allowed visits for one hour during a four-hour period every other Saturday. I went in and met with her. The place was like a cross between an office building and a jail. Institutional cream walls and secretaries behind desks, with potted fake plants in the corners. They put Kareena with me, in a small room with a plastic green desk and two plastic chairs.

She was wearing a blue jump suit, issued by the institution, like a prisoner's, but otherwise she was looking good—alert, sober, eyes not red from substance abuse, lack of sleep, or weeping. She peered at me from her black, triangular glasses and seemed surprised that I should be there.

"You know what's going on out there?" I asked her, after we'd exchanged initial greetings and small talk.

"Yes, we have video."

"It's coming down around our ears, Kareena."

"It's been doing that for some time, I think."

"How long before you get out?"

"They're saying June. But I'll need outpatient follow-up after that."

"Kareena, I'm sorry the way things worked out. I mean, I'm not sorry to have custody of Shannon, but—"

"I know what you mean. I'm sorry, too. But I can't go back."

"Go back—?"

It struck me that she thought I might be wanting to kick start our relationship. That wasn't what I wanted, but I pretended to be following her lead on it.

"I guess you can never go back," I said. "But now something else has come up. Tom and Irene are planning to leave for Alaska. She went out there when Tom was in prison, and now they're wanting to go up there together with Tommy, and Grady and Cassandra Lurie. And—well, they think it might be good if I went along."

Her eyes locked onto mine.

"I told them I have Shannon now, and if I take her with me, you might not like it. Tom said—this was his idea—that maybe you could come too. Not to be with me, but to stay near Shannon."

Those bird-like eyes stayed locked onto mine for a long time. She said nothing for those moments. Then she turned her eyes toward the blank wall and waved a pale, thin hand through the air.

"Jeff, I know things are really going down a rat hole. If you want to know what I think, I think—you should go with Tom and Irene."

I must have gasped. "And Shannon?"

"You should take Shannon with you."

She turned toward me again and quickly added, "You should let me know where you are, so that when I get out—and when I'm ready—"

"I understand," I said. "That means you've got to let me know where you are, too. It shouldn't be hard, Kareena. I've got a million devices. I can be reached anywhere, any time of day or night."

"Then we understand each other," she said.

She seemed strangely calm when she said it. She didn't weep, and her affect was a bit flat. Maybe after I left, she broke down, but how would I know?

I left that three-storey brick building on the east edge of the main drag of town and stepped out onto the street. I punched my umbrella up against the light rainfall. The sky was orange with black cloud cover, rumbling, but no big boomers. But it was springtime, and springtime in Ohio means thunderstorms.

A great hurricane was swirling off the coast of North Carolina too—Hurricane Bruce. It was on track to make landfall in the next couple of days, and then whatever was left of it as it moved northward would send its cold, blowy and wet remnants into our valley.

87

Treasonously Hopeful

We had to make the trek by train. We sold off most of our belongings and wheeled transport before that, so that we'd have as little to carry as possible. I had one suitcase and a big hand basket that I used to carry Shannon's supplies. We had to take an electric taxi to get to the train station in Cleveland, and from there, with a couple of transfers, we were able to get all the way to Vancouver, British Columbia. The train had baby diaper changing stations and a dining car, as well as sleeper compartments, and it was a form of travel I had not experienced before, but liked very well. You could go to sleep in one part of the country and wake up in another, or you could watch the scenery go by: cow pastures in Wisconsin, big blue mountains in Montana, that people kept telling us once had snow caps on them and were much more beautiful that way ...

The rest of the trip to Anchorage was accomplished with a combination of vehicles, from wind cars to electric buses, to horses and buggies we hired out. I accessed a copy of the Anchorage Dispatch on my wearables and started reading a story about a methane blow hole located up near Fairbanks.

It looked like there weren't any safe directions left on the map. It was methane blow holes or hurricanes or fires or creeping coastlines no matter which way you looked. But we had to go somewhere, so we went in the direction that refugees generally took: north. We went together: Grady, Cassandra, their baby, Tom, Irene, Tommy, me and Shannon.

We approached Anchorage from the south, obviously—the land bordering a tidal inlet called Turnagain Arm. So when you looked out, you saw water facing the city, something that is always pleasant to the eye. Not so pleasant was a huge area of tents and tin shacks outside the city skyline, which housed all the refugees. We were determined not to end up there, and we wouldn't have to, if I could buy a house with the money left to me by my mother from my uncle's estate before money lost all its value. I planned to buy a house big enough to hold all of us—three families—and hoped to have enough left over to pay taxes on it for at least ten years. Above the Anchorage cityscape were the Chugach Mountains, which, we were told, once had snow caps, and sometimes did in winters even today. They were large, impressive mountains, etched cleanly against a darkening, reddened sky.

We all felt the wonder of starting a new life.

Down in the Lower Forty-Eight, the primaries kept going forward, but there was no government in Congress since the occupation. Negotiations were underway, but everyone knew that soon there would be no funding, and without funding, services would shut down, more people would be unemployed, markets would panic, and everything would get worse from there.

Wesley Wright was going ballistic on the campaign trail. He'd won a handful of primaries and he could smell victory. He said we needed someone strong, someone willing to take charge, someone to stand up to the protesters. Goya, he said, was a miserable, puling, milquetoast weakling who didn't deserve even to finish out his term and should be immediately impeached.

But who was going to impeach him?

Congress was not in session.

Without funding, the war against the Caliphate would dry up, because we could no longer put soldiers in the field (or the desert). Then we would go home. And as our country broke up into separate states or separate regions, perhaps the terrorists would no longer feel any need to rattle their sabers at us, knowing that we'd never set foot on their land again, never insult them with our religion or our apostasy, our capitalism, our movies, our military hardware and aircraft: never be worth thinking about again.

And as long as I was feeling treasonously hopeful, I also thought that maybe Kareena would reach a point in her outpatient therapy where she felt strong enough to set out for a new land, where she could be a mother, just as mothers had been mothers since the very beginning. But I did not think we would ever be a couple again, and I didn't want it. I was sure Kareena didn't want it, either.

And in the vids, I saw that Nola—Adrienne Raines—was seen as either the wicked traitor who had brought down the American Empire, or as some kind of Joan of Arc who had brought us out of darkness. I only hoped she wouldn't end up the way Joan of Arc did.

As it turned out, I received a long email attachment from her about a month after my arrival in Alaska.

88

That Leaves Us with Baseball

I wasn't hard to track down electronically. I had my email listed on my social media sites, so all Nola—or Adrienne—would have had to do would be to keyword me, and there I'd be.

So when I got an email from "Adrienne" and saw that it contained an attachment, I thought I knew who it was from, and I was eager to read the attachment.

Here's what it said:

Jeff, cheers—
I guess you probably know what's been happening with me lately. You probably wonder why I did what I did. I like to think you might be pleased by it. Well, I surprised myself by what I did. Yet I can see how it came about. To understand it, you'd have to know my whole history. I think to understand anyone, you almost need to know their entire history, and it's impossible to give something that detailed in a small space like an email, or even in a longer space like a book.

Nevertheless, I'm going to try to give you some key points, and then you'll see, I guess.

I come from a family of spies. They didn't call themselves spies, but that's what they were. I told you that my parents were diplomats. Between them they spoke eight languages. I spent my childhood and youth in places like Paris, Berlin, Riyadh and Moscow (where I am now). I ate dinner at the same table with heads of state, prime ministers, cabinet chiefs, translators, scholars, writers, activists, ambassadors, Nobel Prize winners.

I became aware of my parents' 'intelligence' activities when I was a teenager—during that difficult time we all know as the onset of puberty. I remember asking my father if he was a 'spy'. He laughed and said he wasn't. I didn't believe him.

My father was a very handsome man. At about the same time I learned of his secret activities, I also learned that he was habitually unfaithful to my mother. I remember asking him back then, why people were spies, but I think what I really wanted to know was why some men were unfaithful to their wives.

He laughed. "It's all about Mom, apple pie, and baseball," he said.

Mom, apple pie, and baseball. Well, Jeff, apples are contraband for all but the very wealthy, and your mother was killed to protect a state secret. I guess that leaves us with baseball, doesn't it?

My mother looked the other way with regard to my dad's indiscretions. Dad ran with a fast set—famous people, movie stars, people who book the recreational space flights. Some of the people he knew were nothing but rich lowlifes. One of his friends raped me when I was fourteen.

I never told my mother about it. She used to tell me that it was important to hang on to my virginity until my mind had a chance to catch up to my body's rapid maturation. "Don't throw yourself away," she would say. "No one respects you or loves you for it."

I felt like I'd already been thrown away. As I grew older, I patterned myself more after my father than my mother. If it was all about using people and throwing them away, I wanted to be the one to do the using and throwing away. I went to college with the idea of pursuing a career like that of my parents. I guess you knew that story about my taking classes in your town was only a story, nothing to it. I'd already been well-educated (Smith and Columbia, graduate work in international relations).

I guess you know you were an assignment. I've had other assignments before. I did like you, though. You reminded me of Slade Henry, that actor that movie fanatics say looked like that old star from the two-dimensional films, what was his name? James Dean, I think.

As I got older, I began to identify less with my father and more with my mother. She'd been a victim, yet she'd kept something of her humanity, her self-respect, her empathy for others, something that was more than simply the use of her gifts for their own sake or for pleasure. With my father, it was as if that's all that mattered: using your brain to play a game, using your body like a toy. If you cared about anything very deeply, you were some kind of fool. He would just laugh at people who cared deeply about something or believed in something beyond themselves.

In the beginning, when I went to work for the agencies—there are a great many of them, you know, Jeff—I was challenged by it, and I felt I was serving an important cause. National Security. Freedom of the Homeland. Free Markets. But more and more, I would go to dinner parties and listen to middle-aged men talk about their defense contracts, and how they'd bought their seventh or eighth vacation home in Switzerland or the Bahamas or someplace. No one seemed to take seriously the young men and women who were killed in our overseas adventures and conflicts. That was a given, it was something you accepted without a thought.

There were people I knew—people I went to dinner parties with—who were nothing more than drug lords, gangsters, really. I can't say I heard many of them talking about people they'd had killed, but occasionally I did hear them speak of 'expendables'. What were 'expendables', I wondered.

And then when I found out your mother had been killed, I thought, what is this for? Why did this woman die?

I couldn't believe that our nation was more secure because she died.

And then the secret didn't seem worth protecting. And then the secret seemed positively crying out to be exposed.

And so I turned.

I'm a traitor now, or so the story is given out.

I don't think you would think of me that way, would you?

Do our ideas count, Jeff? Yours and mine?

I feel better now than I have in a long time. Wish me luck, Jeff.

Farewell.

Adrienne

At first I was amazed that she had sent me this message over the email, because everything in cyberspace is interceptable. That's why if you say you like green baskets in an email, the next day you'll get fifteen advertisements for green baskets.

But then I thought that there was really nothing in the email that wasn't known in the press already—at least nothing that would be incriminating. She had already admitted leaking the information. The reasons didn't matter to those who felt she was a traitor in need of punishment.

I read her letter over and over. What struck me the most about it was her question, 'Do our ideas count, Jeff? Yours and mine?'

I've thought about that question a long time. I mean, what she was really saying was, do you and I get any say-so about what 'national security' really is, what it means? And what if we see it in a way that it's not usually given to us to see? Are we evil? Are we traitors? Are we un-American? Or are we the real Americans—or the real citizens of the world? Who gets to be right?

I don't know the answer to that. What difference would it make if I did? Who am I? Who listens to Jeff Claymarker?

I guess I could go on, but I feel like I'm coming to the end of my story. Here I am now. I'm an Alaskan. I'm still an American, but who knows how long America will last? What is America anyway? What was it?

You might think, from everything that happened to me, to my family and to my friends, that I hate my country for what it put me through. But I don't hate my country. There is much more to my country than what it did to me. There were beautiful things about it. And I have a daughter now, and I have to try to give her something.

So I ask myself, what is it that was so beautiful?

And then the answer comes to me. It was the land. The land and the dream. I think that's what it was. Whatever it was, or is, it's all I have to give to my daughter, even if it's dying. So I'm going to try to give her that.

I will give her a land. I will give her a dream.

Also by Clarke W. Owens

Son of Yahweh: The Gospels as Novels

COSMIC
EGG
BOOKS

If you prefer to spend your nights with Vampires and
Werewolves rather than the mundane then we publish the books
for you. If your preference is for Dragons and Faeries or Angels
and Demons – we should be your first stop. Perhaps your
perfect partner has artificial skin or comes from another planet –
step right this way. Our curiosity shop contains treasures you
will enjoy unearthing. If your passion is Fantasy (including
magical realism and spiritual fantasy), Horror or Science Fiction
(including Steampunk), Cosmic Egg books will
feed your hunger.